FOUNDATIONS OF
MODERN BIOCHEMISTRY SERIES

Lowell Hager and Finn Wold, editors

ORGANIC CHEMISTRY OF BIOLOGICAL COMPOUNDS*
Robert Barker

INTERMEDIARY METABOLISM AND ITS REGULATION
Joseph Larner

PHYSICAL BIOCHEMISTRY
Kensal Edward Van Holde

MACROMOLECULES: STRUCTURE AND FUNCTION
Finn Wold

SPECIAL TOPICS

BIOCHEMICAL ENDOCRINOLOGY OF THE VERTEBRATES
Earl Frieden and Harry Lipner

THE BIOCHEMISTRY OF GREEN PLANTS
David W. Krogmann

EXPERIMENTAL TECHNIQUES IN BIOCHEMISTRY
J. M. Brewer, A. J. Pesce, and R. B. Ashworth

*Published jointly in Prentice-Hall's *Foundations of Modern Organic Chemistry Series*.

THE
BIOCHEMISTRY
OF
GREEN PLANTS

THE BIOCHEMISTRY OF GREEN PLANTS

DAVID W. KROGMANN

Professor of Biochemistry
Purdue University

Prentice-Hall, Inc., Englewood Cliffs, New Jersey

Library of Congress Cataloging in Publication Data

KROGMANN, DAVID W.
    The biochemistry of green plants.

    (Foundations of modern biochemistry series)
    Bibliography: p.
    1. Botanical chemistry.   I. Title.
QK861.K7   581.1'92   73-7637
ISBN 0-13-076463-9
ISBN 0-13-076455-8 (pbk.)

Printed in the United States of America.

10 9 8 7 6 5 4 3 2 1

PRENTICE-HALL INTERNATIONAL, INC., *London*
PRENTICE-HALL OF AUSTRALIA, PTY. LTD., *Sydney*
PRENTICE-HALL OF CANADA, LTD., *Toronto*
PRENTICE-HALL OF INDIA PRIVATE LIMITED, *New Delhi*
PRENTICE-HALL OF JAPAN, INC., *Tokyo*

The poem "The Termite," which appears on page 69, is from *Verses from 1929 On* by Ogden Nash and is used with the permission of Little, Brown and Co., The Estate of Ogden Nash, and J. M. Dent. Copyright, 1942, by The Curtis Publishing Company.

The poem "Unseen and Seen" by Dorothy Long, which appears on page 199, is from *Perspectives in Biology and Medicine*, Vol. 15, Spring 1972, and is used with the permission of The University of Chicago Press.

To RETT, MICHELLE, PAT, and PAUL
and to BARNEY AXELROD

CONTENTS

PREFACE

This text is directed to the needs of students who are interested in the biochemistry of green plants – not only those students specializing in biochemistry but also those in botany, plant physiology, agronomy, plant pathology, horticulture, forestry, and ecology. The contents can be covered in a one-semester course and are by no means comprehensive. Since a course in general biochemistry is a *sine qua non* for any advanced study in the life sciences, such an instruction is taken for granted as a prerequisite to the more specialized areas covered in this text. Since botanical terms, plant structures, and even the generic names for plant species are often of use but rarely remembered, a glossary has been appended. At the end of most chapters a few general references are given to lead the reader to more detailed summaries; footnotes are cited in the text by number and collected as references at the back of the book to provide leads into the research literature.

There is little point in cataloging all of the biochemical processes common to all living things and noting their occurrence in plants. Usually these general processes were first worked out with mammalian or bacterial preparations which are inherently easier to do biochemistry with. The subsequent discovery of these processes in plants has been confirmatory and underscores the general metabolic unity of life. Hence the content of this book is limited to biochemical processes which are especially germane to plant tissues and, even then, not all of the strictly botanical chemistry has been included. There is a vast area of natural products chemistry embracing the "secondary metabolites" of plants which include alkaloids, phytosterols, many pigments, and so on. The metabolism of these compounds is being worked out. The physiological role, if any, for the "secondary metabolites" is largely obscure. Since there are as yet no general principles or functional generalizations relating to this area — only a very large number of facts — the reader is best served by going to the research literature for needed details in the "secondary metabolite" area. This book makes no pretense at completeness in covering all areas of plant metabolism of "primary metabolites." Some problems were selected where the vitality of contemporary research and the uniqueness of substance pointed to such areas as critical to an appreciation of the biochemistry of plant materials. Other areas have been omitted, since work with plant tissues has added very little to the chapters which appear in the four basic volumes of the *Foundations of Modern Biochemistry* series.

The special fascination in understanding the life of green plants in chemical terms can only become evident from studying details such as those that follow. The general tendency toward mammalian chauvinism and the spectacular successes in understanding *Escherichia coli* and viruses often push green plants back toward obscurity. Still, green plants have provided the earth's atmosphere with oxygen thereby paving the way for great advances in the evolution of life; green plants carry out photosynthesis on an annual scale which exceeds by orders of magnitude man's greatest efforts in chemical processing; and green plants are the basis for all other contemporary heterotrophic forms of life. How can one understand the world without understanding plants? In a more practical sense, the "Green Revolution" of modern agriculture has a great deal of plant biochemistry at its base. While society puzzles in anguish at the prospect of the genetic engineering of man, one suspects that the genetic engineering of plants will be accomplished quietly and *pro bono publico*, if it hasn't happened already. The development of high lysine corn is an example of a straightforward application of biochemistry to plant tissue which is revolutionizing the nutritional value of man's food. A major food grain which had previously been deficient in supplying the essential amino acid lysine to the mammalian diet has been genetically altered to supply this amino acid since biochemical analyses led to the recognition and correction of this deficiency. Thus on both theoretical and practical grounds, plant biochemistry deserves attention.

Thanks are due to many students and colleagues who gave suggestions and encouragement to this writing. I am especially grateful to Beverly Burwell who was most generous in her help with library materials and reference verification, to Jennie Johannesen for her skillful work in preparing the manuscript, and to Loretta Krogmann and Rita Barr for their many readings and suggestions.

DAVID W. KROGMANN

1 || **INTRODUCTION**

At the outset of a study of plant biochemistry, one must ask if there is any unique province for the biochemistry of plant cells. Much of the metabolism of plant tissue is identical to that found in animal or bacterial cells. The biochemical activities of plants are clearly unique with respect to

pigments
photosynthesis
cell walls
chloroplasts
photoregulation
metabolism of certain inorganic elements
phytohormones
secondary metabolites

The tendency in the discussions that follow is to concentrate on biochemical processes which are general to most plant tissue and to avoid that which is peculiarly characteristic of a single species or group.

1

The architecture of the plant cell includes the

nucleus
mitochondria
endoplasmic reticulum bearing ribosomes
Golgi apparatus
microbodies
plasma membrane
fat droplets
carbohydrate granules
protein bodies

which are generally similar to structures found in animal cells. In addition, the plant cell has

a rigid cell wall
a vacuole
chloroplasts
glyoxysomes

1.1 EXPERIMENTAL CONSIDERATIONS UNIQUE TO PLANTS

The unique difficulties in doing biochemical experiments with plants are formidable and arise from the special character of plant cell structure.

1. Much of the bulk of fresh plant material is water — 90% of a spinach leaf is water, so one is working with material that is inherently dilute.

2. The cell wall makes up one-fourth to one-half of the dry weight of plant material. The wall is practically inert in a biochemical sense and is frequently so tough in a physical sense that breaking the wall often involves breaking the structured contents of the cell. Many marine algae are essentially unbreakable. A variety of mechanical devices are available for homogenizing plant tissue so that one can usually find circumstances that allow isolation of a particular organelle. Enzymatic digestion of the plant cell wall can be used to produce protoplasts which can then be broken under very gentle conditions.

3. In addition to wall material, some plants may accumulate various substances which, when the cell is broken, form a gel that enmeshes all the cell contents in an unfractionatable mess. Parsley and red algae show this property. In principle, one should be able to find an enzyme to hydrolyse this polysaccharide but in practice such enzymes are often hard to obtain.

4. Most plants contain phenolic compounds that oxidize and polymerize when the cell is broken open. The polymers are frequently acidic, frequently bind irreversibly to proteins, and are always colored — yellow

to dark brown — which makes spectrophotometric assays difficult. The oxidation of these compounds can be minimized by grinding the plant tissue with a reducing agent like ascorbic acid or glutathione and by using a metal chelator to suppress some of the metal or metalloenzyme catalysed oxidations. Some of the oxidized polymeric phenols can be absorbed on an added synthetic polymer like polyvinyl pyrolidone, nylon powder, or even ion exchange resins.

5. The plant cell vacuole often serves as a cess pool in this creature which has no excretory system. The frequently noxious material accumulated in the vacuole is mixed with the protoplasm when the cell is broken open and may denature or inhibit the enzymatic machinery. In the case of acid accumulating Crassulacean plants, this problem has been overcome by infiltrating the tissue with alkaline buffer before homogenizing.

6. The metabolic rates in many plant tissues are relatively low compared to bacterial and animal cells (the problem of dilution with water and cell wall plus a slower rate of cell replication). Thus, most plant metabolism studies begin with isotope measurements.

7. The problem of bacterial contamination of many tissues is quite a danger when studying low-rate phenomena in tissue homogenates — many experiments on amino acid incorporation into chloroplasts and plant mitochondria went down the drain when bacterial contamination was properly indentified. The use of axenic cultures of algae or of higher plant cells in tissue culture is one obvious way around this problem. Otherwise, careful controls must be used to subtract or correct for the misleading effects of bacterial contamination.

To illustrate the dilution problem, consider that of the one-half of the dry weight (i.e., 5% of the fresh weight) which is not cell wall, roughly half is soluble cytoplasmic constituents and half is particulate material. For either fraction, roughly half is protein. To obtain one gram of unfractionated leaf cytoplasmic protein, something over 80 grams of fresh leaves are required.

The size and number of cellular constituents are indicated in Table 1.1. In 1947, one guessed that a cell might contain 1,000 different kinds of enzymes.

TABLE 1.1   SIZES AND NUMBERS OF PLANT CELL STRUCTURES

	Diameter in Angstroms	*Number per Cell*
Nucleus	$5-20 \times 10^3$	1
Chloroplasts	$4-10 \times 10^3$	50–200
Mitochondria	$1-5 \times 10^3$	500–2,000
Microbodies	$2-15 \times 10^3$	300–3,000
Ribosomes	250	$5-50 \times 10^5$
Enzymes	20–100	$5-50 \times 10^8$

Bonner[1] has revised the estimate by one order of magnitude to 10,000 kinds of enzymes and adds another 100 kinds of non-enzymic proteins per cell. The non-enzymic proteins — structure proteins and nitrogenous reserves — are frequently much more abundant than individual enzymes, so that in enzyme isolation one usually needs very large quantities of plant material to obtain a reasonable amount of pure enzyme.

1.2   THE MAGNITUDE OF PLANT PROCESSES

Before plunging into the detailed substance of plant biochemistry, consider briefly the magnitude of the total process of plant metabolism. M. Kamen is fond of decrying the twin vices of "mammalian chauvinism and temporal solipsism" which restrict enthusiasm for plants and their works.[2,3] Some gee-whiz numbers might stimulate a look beyond the more conventional horizons. Approximately 100 billion (10^{11}) tons of carbon dioxide are consumed by photosynthesis per year.[4,5] This activity may be compared to the 4 billion (4×10^9) tons of fossil fuel consumed or the 400 million (4×10^8) tons of steel produced per year in the last decade.[6] In terms of magnitude of per annum activity, green plants handle about 250 times as much material as produced by the steel industry. The total net deposit of photosynthesis in terms of fossil fuel reserve on this planet is estimated at 10^{13} tons.

The temporal magnitude of photosynthesis is only vaguely reflected in the fossil fuel figures, and recent geochemical studies indicate that photosynthesis has been going on for a long time. Barghoorn has made interesting attempts at establishing chemical evidence for the antiquity of life beyond the age of obvious fossil deposition. The age of the earth is currently set at 4.9×10^9 years by radioactive uranium-thorium dating. Barghoorn took carbonaceous deposits that could be accurately dated by radiochemical measurements and he observed what appear to be fossil algae in electron micrographs of these samples.[7] In order to obtain chemical evidence, these rocks were carefully surface cleaned to remove contemporary contaminants, pulverized, extracted, and the extract hydrolyzed in hydrochloric acid to release amino acids from possible peptides.[8] The extract was then run through an amino acid analyzer, with the results as shown in Table 1.2 and Figure 1.1. Even without peptide hydrolysis, the samples

TABLE 1.2   FREE AMINO ACID CONTENT OF CHERT

Specimen	Age $\times 10^9$ Years	Glycine nmoles/10 gm Sample
Bitter Springs	1.0	18.1
Gunflint	1.9	9.8
Fig Tree	3.1	5.9

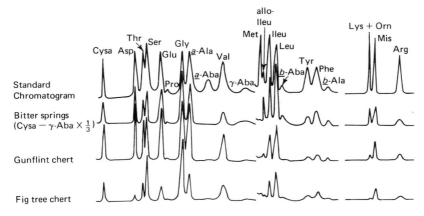

Figure 1.1   Amino acid analyzer chromatograms of the acid hydrolysates of three Precambrian cherts. [From Schopf, J. W., K. A. Kenvolden, and E. S. Barghoorn, *Proc. Natl. Acad. Sci. USA,* 59, 639 (1968).]

are seen to contain free amino acids whose concentration is related to the antiquity of the sample. The acid hydrolysates from the rock extract showed amino acid analyzer patterns of decreasing complexity with increasing age, presumably because of the relative instability of different amino acids. Identity of eleven amino acids was checked by derivitization and vapor-phase chromatography. Although this kind of data does not rule out an abiogenic origin of the compounds, it is an interesting probe at the biochemical information available in fossil material.

Another more frequently cited evidence of ancient plant life is the appearance of porphyrins and certain hydrocarbons in oil shale. The hydrocarbons pristane and phytane are particularly interesting since they can be detected with great sensitivity and unambiguous accuracy in the mass spectrometer. These compounds are presumed to arise from the degradation of chlorophyll via the reactions in Figure 1.2. Pristane and phytane have been identified in the Gunflint chert and in the Fig Tree material dating chlorophyll back to more than 3 billion years.

The Gunflint deposits show structures that may be fossil blue-green algae and other evidence suggests that this material was deposited at the beginning of atmospheric oxygen production. On taxonomic grounds the blue-green algae are the most primitive of oxygen producing plants. Contemporary blue-green algae show little respiratory oxygen consumption and no activity in the mixed function oxidase reactions where oxygen is used for biosynthetic purposes. It is reasonable that organisms responsible for the appearance of oxygen in the atmosphere would not need oxygen for growth.

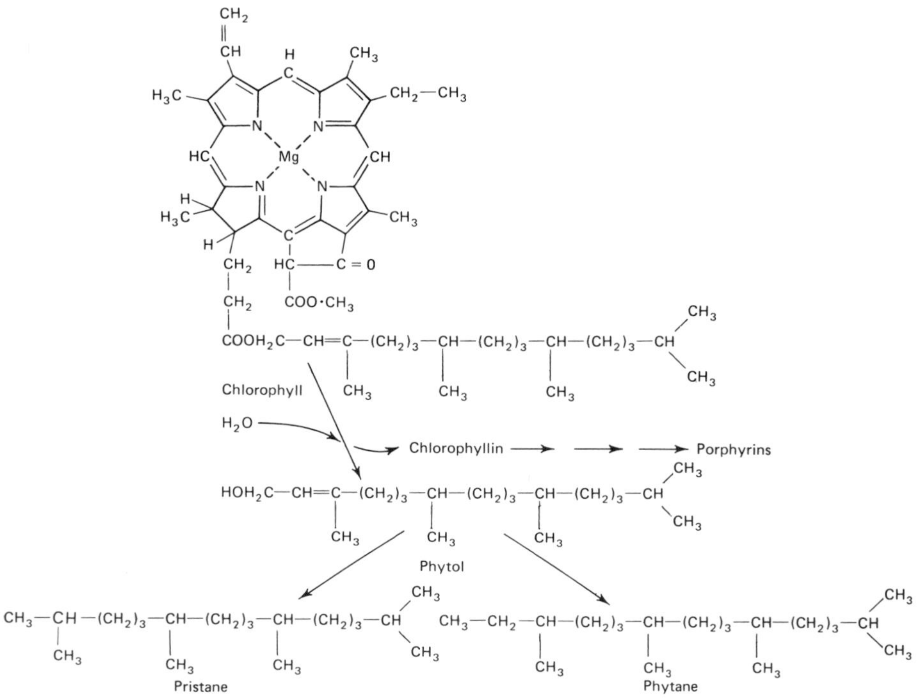

Figure 1.2 The degradation of chlorophyll to stable hydrocarbons of characteristic structure.

1.3 THE EVOLUTION OF PLANT CONSTITUENTS

With the consolidation of descriptive knowledge of biochemistry, it has become possible to try to distinguish ancient from more modern biochemical processes or to seek this history in the structure of macromolecules. A few examples are of interest at this point, but the theme of biochemical evolution will turn up again and again.

The ability to synthesize certain complex metabolites is an indication of the position of the plant tissue on the evolutionary scale. Sterols and certain unsaturated fatty acids are found in red algae and higher plants but not in bacteria and only occasionally in blue-green algae.[9] This is consistent with the theory that blue-green algae are responsible for the introduction of oxygen in the earth's atmosphere since these compounds are formed through oxygenase type reactions. These processes could only evolve after atmospheric oxygen became available.

The structure of complex macromolecules contains detailed information about the evolutionary position of the organism and amino acid sequencing has opened a whole new world of taxonomic and evolutionary comparisons. The ferredoxin molecule is an interesting case in point. This is a low molecular weight protein found in certain anaerobic heterotrophic bacteria and in photosynthetic autotrophs. Since ferredoxin is a small protein and is readily purified, it is admirably suited for sequence work. A ferredoxin sequence is shown in Figure 1.3. A comparison of the two halves of the ferredoxin isolated from the heterotrophic bacterium *Clostridium pasteurianum* indicates that this molecule arose by the doubling of a relatively short gene.[10]

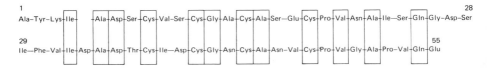

1

Ala–Tyr–Lys–Ile– –Ala–Asp–Ser–Cys–Val–Ser–Cys–Gly–Ala–Cys–Ala–Ser–Glu–Cys–Pro–Val–Asn–Ala–Ile—Ser–Gln–Gly–Asp–Ser

28

29

Ile—Phe–Val–Ile–Asp–Ala–Asp–Thr–Cys–Ile—Asp–Cys–Gly–Asn–Cys–Ala–Asn–Val–Cys–Pro–Val–Gly–Ala–Pro–Val–Gln–Glu

55

Figure 1.3   Amino acid sequence of ferredoxin from *Clostridium pasteurianum*. Twelve of the first twenty-eight residues are found in corresponding positions in the latter half of the sequence from residue 29 to 55. A detailed comparison of the nucleic acid codons responsible for positioning each of these amino acids suggests that most of the variations between the two halves resulted from a change of only one nucleotide per codon.

The ferredoxin from the photosynthetic bacterium *Chlorobium* has been isolated and appears similar to Clostridial ferredoxin in size and amino acid composition. However, the ferredoxin from the photosynthetic bacterium *Chromatium* is larger — 81 residues instead of the 53 to 57 residue lengths found for other bacterial ferredoxins. Ferredoxins from blue-green algae, green algae, and from a number of higher plants show a characteristic size of between 97 to 105 residues. The clear indication of gene doubling in the short bacterial ferredoxins prompts the hypothesis that the gene tripled in *Chromatium* and quadrupled in algae and higher plants.[11] More sequences of bacterial and algal ferredoxins are needed to provide convergent evidence for this suggestion.

Amino acid sequences for higher plant ferredoxin are rapidly becoming available and four sequences are shown in Figure 1.4. The sequences are for ferredoxins from *Leucaena glauca* — a small, leguminous tree, spinach, alfalfa, and *Scenedesmus* — a green alga. In all but the first line, only the positions of variance are shown so that even a cursory glance indicates that the molecules are very similar.

In the *L. glauca* sequence, there are four positions where two amino acid residues are indicated for the same location (6, 12, 33, and 96).[12] Positions 6 and 12 occur within the same tryptic peptide fragment, and this peptide is found in two forms (6-Leu, 12-Pro and 6-Val, 12-Ala) indicating microheterogeneity

Positions: 1 ⋯ Leu ⋯ 10 ⋯ Pro ⋯ 20

	1	2	3	4	5	6	7	8	9	10	11	12	13	14	15	16	17	18	19	20	21	22	23
L. glauca	Ala	Phe	Lys	Val	Lys	Val	Leu	Thr	Pro	Asp	Gly	Ala	Lys	Glu	Phe	Glu	Cys	Pro	Asp	Asp	Val	Tyr	Ile
Spinach	[Ala]	Tyr			Thr	Leu	Val		Thr				Asn	Val		Gln							
Alfalfa	Ala	Tyr			Lys	Leu	Val		Glu				Thr	Gln		Glu							
Scenedesmus	Ala	Tyr			Thr	Leu	Lys		Ser		Asp	Gln	Thr	Ile		Glu					Thr		

Positions: 53 ⋯ 60 ⋯ 70

	53	54	55	56	57	58	59	60	61	62	63	64	65	66	67	68	69	70	71	72	73	74	75	76	77
L. glauca	Glu	Gly	Asp	Leu	Asp	Gln	Ser	Asp	Gln	Ser	Phe	Leu	Asp	Asp	Glu	Gln	Ile	Glu	Glu	Gly	Trp	Val	Leu	Thr	Cys
Spinach	Thr		Ser		Asn			Asp							Asp			Asp							
Alfalfa	[Ala]		Glu	Val	Asp		Ser		Gly						Asp			Glu							
Scenedesmus	Ala		Thr	Val	Asp		Ser		Gln						Ser	Met	Asp	Gly			Phe				

Figure 1.4   A comparison of amino-acid sequences of ferredoxins from green plants.

within this protein. Since this heterogeneity in sequence is found in a protein sample isolated from a single tree, it is not because of subspecies variation among plants.[13] These sequence ambiguities might result from polyploidy — a duplicate set of chromosomes or genes are present in the cell but are not exact duplicates as a result of independent mutations, so they code for slightly different proteins.

One can quantitate the comparison of these ferredoxins by noting that the higher plant proteins differ among themselves in from 18 to 21 positions while the algal ferredoxin differs in 28 or 29 positions from the higher plant ferredoxins. Since an amino acid substitution might be the result of 1, 2, or 3 nucleotide changes, the degree of difference between proteins can be refined by considering the nucleotide triplet that constitutes the codon for each residue. The degree of difference in nucleotide sequence can be expressed as the minimum base difference per codon for the proteins to be compared. Thus, the following comparisons;

L. glauca/spinach	MBDC = 0.30	(This agrees with the classical taxonomic positions of these
L. glauca/alfalfa	MBDC = 0.26	plants in which *L. glauca* is more closely related to alfalfa
spinach/alfalfa	MBDC = 0.28	than to spinach.)
Scenedesmus/spinach	MBDC = 0.41	(Obviously, algae are more remotely related to higher plants
Scenedesmus/alfalfa	MBDC = 0.42	than higher plants are related to one another.)

Comparison of the amino acid sequences and attendant codon sequences of the ferredoxins from *Scenedesmus* and spinach suggests a mechanism for the evolution at residues 51 to 53. A frame shift mutation as illustrated in Figure 1.5 could have occurred by dropping a deoxyguanilic from the Val codon at position 51 and by adding a deoxyadenylic at the Ala codon at position 53.

```
                    30                    Asp                      40                                              50
Leu Asp Gln Ala Glu Glu Leu Gly Ile Glu Leu Pro Tyr Ser Cys Arg Ala Gly Ser Cys Ser Ser Cys Ala Gly Lys Leu Val
                        Glu     Ile                                                                              
                [Ala]   Lys     Met Asp                                                                    Lys
                His     Glu     Ile Val                                                                Val [Ala]
                Ala     Ala     Leu Asp                                  Ala                           Val Glu


      80                              90                          Gly
Ala Ala Tyr Pro Arg Ser Asp Val Val Ile Glu Thr His Lys Glu Glu Glu Leu Thr Ala
[   ]           Val             Thr                                      Ala
Val         Ala Lys         Thr                                         Ala
Val         Pro Thr     Cys Thr  Ala                    Asp Phe
```

The evidence for gene duplication or quadruplication in the higher plant ferredoxins is largely obliterated from the sequences since a lot of mutations may have occurred within the subsections. It is interesting to note that the four pairs of Ala-Ala in the alfalfa protein start segments of 26, 25, 26, and 20 residues in length with most of the variable positions clustered near the amino terminal end of these segments.

```
              51            53
          Lys – Val – Glu – Ala – Gly –          Scenedesmus

          AAG  GTT  GAA  GCT  GGC
              ↳delete

                              insert
          AAG  TTG  AAG  ↲ACT  GGC

          Lys – Leu – Lys – Thr – Gly –          Spinach
```

Figure 1.5   A possible frame shift mutation in the evolution of ferredoxin. Deletion of a nucleotide in codon 51 and insertion of a nucleotide in codon 53 allows the preservation of the amino acid sequence beyond position 53.

The prime example of the contribution of amino acid sequence to evolutionary and taxonomic comparisons has come from comparing sequences of the protein cytochrome c. Although most of the sequences have been done for mammalian cytochromes, the few sequences for plant cytochromes permit interesting inferences. Kamen has obtained a sequence for cytochrome c_2 of the photosynthetic bacterium *Rhodospirillum rubrum*.[14] First, there is only slight homology to mammalian cytochrome c so this is a sequence that is largely altered over the early stage of evolution. In addition, the sequence of *R. rubrum* cytochrome c of 112 amino acids shows a 13 to 20 amino acid repeating fragment of close internal homology, which again suggests that the structural gene coding this protein arose through partial gene doubling.

Boulter and his associates have sequenced cytochrome c from a number of higher plant species and have begun the construction of a phylogenetic tree which promises considerably greater accuracy than the often incomplete fossil record might afford.[15] Figure 1.6 illustrates some of the results of this work.

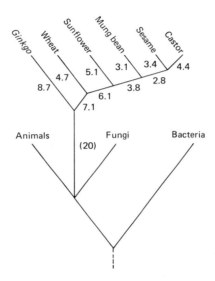

Figure 1.6   A phylogenetic tree relating higher plant cytochromes by the "ancestral sequence method" for analyzing amino acid sequences. [From Ramshaw, J. A. M., M. Richardson, and D. Boulter, *J. Eur. Biochem.*, **23**, 475 (1971).]

Given sufficient data, sequence studies should provide a very precise system of taxonomy. The most exciting possibilities center on the bacteria, algae, and protozoa — organisms in which morphology of the cell and the usual phenotypic comparisons do not allow much inference concerning species interrelations or evolution. The ferredoxin and cytochrome comparisons illustrate the possibility of a much better understanding of geneology of microorganisms. A comparison of the cytochromes in photosynthesis and respiration as illustrated by Kamen's work may clarify the evolution of these two processes.

The comparison of nucleotide sequences in the genes from various species is the fundamental level of biochemical comparison, but chemical sequencing of nucleic acid is a formidable task. Base composition data is of little value as seen from the lack of difference in base content of DNA from the primitive ferns to the more advanced angiosperms.[16] It is possible to make physical measurements of the degree of similarity of nucleic acid sequences in various species. A double-stranded DNA molecule has a characteristic melting point — T_m — or temperature at which the duplex separates into single strands. The melting temperature of the duplex is determined principally by the complementarity of base pairing and by the relative numbers of GC versus AT pairs. A natural duplex should have perfect complementarity since one strand was formed by complementary base pairing to the other strand. The T_m is then the result of the proportion and location of the GC base pairs which bond a bit more strongly than the AT pairs. A solution of melted DNA can be slowly cooled in a fashion that allows the duplexes to reform, a process which is called annealing. In order to compare DNA from different species, one melts the homologous duplexes from each species, then mixes the single strands and anneals them to form heterologous duplexes containing one strand from each of the species. The imperfect base pairing in the heterologous duplexes results in a lowering of thermal stability and a consequent lowering of the melting point. A $1°$ lowering of the melting point is approximately equivalent to a 1% change in the base sequence. When barley and wheat DNA are treated in this way, one finds a significant lowering of the barley-barley homologous duplex T_m in the barley-wheat heterologous duplex. Alternately, one can measure the degree of difference in base sequences between various plant species and by measuring annealing in isotope labelled DNAs from different species and by hybridization-competition develop precise quantitative measures of differences in genomes.[17]

2 ‖ HEXOSE BREAKDOWN

      This section might well begin with a brief review of the glycolytic sequence which hopefully is very familiar. Figure 2.1 summarizes the reaction sequence. Glycolysis obviously occurs in plants – the enzymes are recognizable in plant tissue homogenates and complete glycolysis of glucose to pyruvate has been observed with plant extracts. From a physiological point of view it is obvious that plants should glycolyse reserve hexose in the dark to provide energy for cellular maintenance at night. The points about glycolysis to be emphasized in the paragraphs that follow are reversibility, regulatory mechanisms, and the unique properties of individual glycolytic enzymes in plant material.

2.1 REVERSIBILITY OF GLYCOLYSIS

Conceding that the Embden-Myerhoff glycolytic reactions go readily from glucose to pyruvate, some plant tissues must be able to reverse the process. During the germination of fatty seeds, the bulk of the carbon reserve exists as

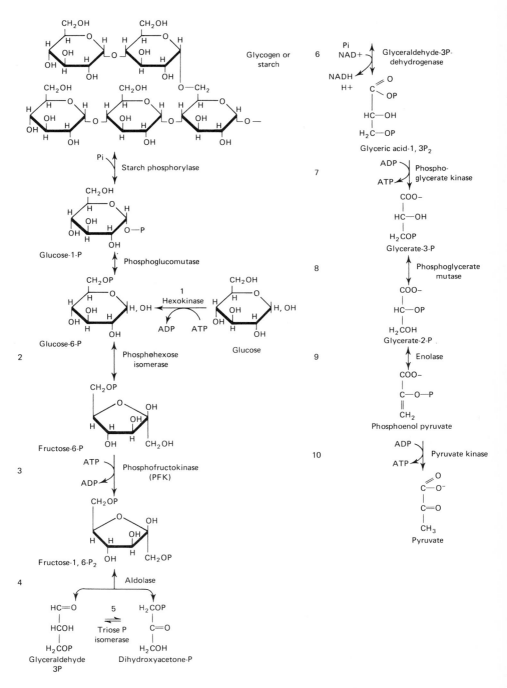

Figure 2.1   The Embden–Meyerhof pathway for glycolysis of hexose.

lipid which must be converted to carbohydrate in order to supply the hexose required by the growing plant. Before the germinating seed can begin photosynthesis, it must send a shoot with lots of hexose cell wall polymers up a considerable distance through the soil. The hexose is supplied by a reversal of glycolysis. The glycolytic steps that are difficult to reverse by the enzymes of conventional glycolysis are pyruvate kinase (10), phosphofructokinase (3), and hexokinase (1). These reactions are provided with bypass routes. Acetyl CoA from lipid degradation is fed in via the glyoxylate cycle as oxaloacetate which is converted directly to phosphoenolpyruvate

$$\text{Oxaloacetate} + \text{ATP} \rightleftharpoons \text{phosphoenolpyruvate} + CO_2 + \text{ADP}$$

bypassing the unfavorable PEP kinase step (10). The glycolytic reactions are then freely reversible back to the level of fructose 1,6 diphosphate and here a fructose 1,6 diphosphate phosphatase is available.

$$H_2O + \text{fructose 1,6 diphosphate} \longrightarrow \text{fructose 6 phosphate} + \text{Pi}$$

The initial step of glycolysis catalyzed by hexokinase is also an irreversible step. This irreversible step could be circumvented by another phosphatase, but this may hardly be necessary, since most hexose metabolism involves the phosphorylated sugar.

Korb and Beevers have studied the overall reversal of glycolysis — gluconeogenesis in castor bean endosperm.[1] This tissue catalyzes a massive conversion of fat to carbohydrate when the castor bean is germinated. Glucogenesis occurs in a different cellular compartment from glycolysis but regulation seems to occur at the same steps for both processes — the fructose 6 phosphate, fructose diphosphate step, and at the phosphoenolypyruvate level.

2.2   REGULATION OF GLYCOLYSIS

Control of enzyme activity via feedback regulation and allosteric mechanisms is now a topic of great interest. The data on regulation of glycolysis in plant tissue is fragmentary. As in animal tissue, phosphofructokinase appears to be a control point. Phosphofructokinase has been purified from carrot tissue and it is inhibited by ATP and citrate, products that might accumulate from excessive glycolysis.[2] Unlike the mammalian enzyme, ATP inhibition is not reversed by AMP and ADP. High levels of inorganic phosphate reverse the ATP inhibition but one must regard phosphate as a dubious regulator of plant metabolism since the phosphate concentration in plant tissue is usually very high — several orders of magnitude higher than the nucleotide coenzymes. It may be that much of this phosphate is in the vacuole or some other compartment, so that fluctuations in a much smaller pool in the protoplasm allow phosphate concentration in a regulatory role. This, like many presumed regulatory effects, needs proof for the *in vivo* state. In another study, phosphofructokinase regulation has been examined in the Brussels sprout,[3] which superficially seems an amusing choice of plant tissue but is in fact a tissue of unique advantage. The innermost tissue is embryonic, but the outermost leaf is mature to senescent. Thus one can simply peel off the life history or temporal stages of leaf development. Phosphofructokinase shows different regulatory properties depending on the age of the tissue. The embryonic tissue has an enzyme that is most sensitive to possible changes in the cellular environment since it is inhibited by ATP and citrate and activated by fructose 6 phosphate and inorganic phosphate. The older tissue has a relatively insensitive enzyme. It would be interesting to isolate these two enzymes and look for physical and chemical differences in structure. The phosphofructokinase in pea seedlings is inhibited by phosphoenolpyruvate, 6 phosphogluconate, and either 2 or 3 phosphoglycerate. This product inhibition can be reversed by high concentrations of fructose 6 phosphate and magnesium ion or by rather modest concentrations of ATP.[4] As in the earlier cited case of the carrot enzyme, these studies are incomplete until one is assured that the regulatory agents show concentration changes *in vivo* that are consistent with their postulated regulatory role.

Fructose 1,6 diphosphate phosphatase is another prime candidate for regulation in view of its unique role in glucogenesis. There is suggestive evidence that the synthesis of fructose diphosphate phosphatase is repressed in germinating wheat embryos by glucose and glycerol (catabolite repression?).[5] The enzyme activity is sensitive to AMP inhibition − if the embryo had high AMP, ATP would presumably be low, so the tissue should slow down its conversion of carbon into hexose and use some to make ATP. Scala et al. have examined the fructose 1,6 diphosphate phosphatase from bean leaves and found that this enzyme is inhibited by fructose 1 phosphate and both ATP and AMP.[6] The fructose 1,6 diphosphate phosphatase was found to increase on germination of castor bean and this increase is due to *de novo* synthesis of a new isozyme. The enzyme induction was attributed to action of the plant hormone gibberellic acid.[7]

2.3   ALDOLASE

The unique properties of two other glycolytic enzymes from plant tissue have been examined in some detail. One of these enzymes is adolase, which occurs in various tissues as one of two distinct types of enzyme.[8] The initially recognized distinctions came from comparing the aldolase of muscle with the aldolase from yeast or bacteria. Table 2.1 shows some of the contrasting characteristics of these two types of enzymes. Early studies of plant aldolases were confusing, but

Gibbs[9] has lent some clarity to the situation. In higher plants, both green leaves and seeds have Class I aldolase. In *Chlorella* and *Euglena*, Class II aldolase is found when growth is either photosynthetic or heterotrophic. Blue-green algae have Class II enzyme as would be expected from their other similarities to bacteria.[10] The transitional species appears to be the alga *Chlamydomonas*[11] which, when grown on acetate in the light, has a Class II aldolase like a heterotrophic bacterium, but when grown autotrophically on CO_2 in the light, it has a Class I aldolase like higher plants. Criddle refined the observation on higher plant material by isolating aldolase from spinach chloroplasts, and he found this to be a Class I enzyme.[12] Horecker has purified and partially characterized the fructose diphosphate aldolase from spinach leaves and measured a molecular weight of 120,000 with four subunits of 30,000 each.[13] There is an indication that the subunits are of two types, as is the case in the muscle enzyme. Takeo developed a specific stain for aldolase and thus was able to look for isozymes with disc gel electrophoresis. In surveying spinach, carrot, and radish, he found two isozymes, one form in the seed and a second form, which becomes the predominant one, in leaves.[14]

TABLE 2.1 COMPARISON OF THE PROPERTIES OF THE MAJOR TYPES OF ALDOLASES

Tissue of Origin	Class I Mammalian Muscle	Class II Yeast and Bacteria
Inhibition of activity by metal chelators	–	+
Potassium ion stimulation	–	+
Sulfhydryl requirement	–	+
Molecular weight	146,000	70,000

Thus in the blue-green algae the enzyme is like that found in bacteria, and this type of enzyme persists in certain of the green algae – *Chlorella* and *Euglena* – despite the evolution to a true nucleus (eucaryotic) and despite the many other similarities to higher plants. In *Chlamydomonas*, either type of enzyme can be synthesized – Class II for heterotrophic growth and Class I for autotrophic growth. In the mature higher plant, there is only Class I aldolase functioning in both glycolysis and photosynthesis although a different type of aldolase is present in seeds.

2.4 GLYCERALDEHYDE PHOSPHATE DEHYDROGENASE

Glyceraldehyde phosphate dehydrogenase (triose phosphate dehydrogenase, TPD) is the second major enzyme shared by glycolysis and photosynthetic carbon fixation that has demanded much attention. This enzyme, when isolated from etiolated tissue, uses NAD specifically as its coenzyme in what must be a largely glycolytic degradation of hexose (step 6 in the diagram of glycolysis) — oxidizing glyceraldehyde 3 phosphate to 1,3 diphosphoglyceric acid. In green tissue, activities linked to both NAD and NADP are found.[15,16] It is assumed that the NADP-linked activity is associated with photosynthetic production of hexose through reduction of 1,3 diphosphoglyceric acid to 3 phosphoglyceralde-hyde. This is consistent with the general beliefs that:

1. NADPH + H^+ is the reducing agent generated by the illuminated chloroplast.

2. NADP is the coenzyme of reductive biosynthesis while NAD is the coenzyme of oxidative, energy generating metabolism.

In the green leaf, the NADP-linked glyceraldehyde phosphate dehydrogenase is localized exclusively in the chloroplast, while the NAD-linked enzyme is found in both the cytoplasm and the chloroplast.[17,18] It is clear that the NADP-linked glyceraldehyde phosphate dehydrogenase activity of higher plants is induced by a phytochrome mediated activation.[19,20] Thus the NADP-linked enzyme can be made to appear without chlorophyll synthesis by red light (the red light induction is reversed by far red light), which converts one form of phytochrome to the other. Since the phytochrome activation also results in an increased synthesis of protochlorophyll, this response seems to be preparing several, but not all, chloroplast components for the onset of photosynthesis. The above experiments were done with bean leaves, but the system for NADP-linked glyceraldehyde phosphate dehydrogenase in pea leaves may be different. Ziegler and Ziegler have found a chlorophyll-less mutant of pea that shows an appearance of NADP-linked enzyme in response to light.[21] It is not clear that this mutant is a point loss of chlorophyll only or an ablation of much of the phytochrome controlled genome. Melandri et al. found that the light induced appearance of the NADP-linked glyceraldehyde phosphate dehydrogenase in peas was not blocked by inhibitors of protein synthesis, but no evidence could be obtained for activation by enzyme conversion or by allosteric effectors.[22]

The Zieglers have made another observation on the glyceraldehyde phosphate dehydrogenase of green tissue that demands exploration. They have found a rhythmic change in the NAD- and NADP-linked enzyme activities related to day-night cycles. The Zieglers have observed that the NADP-linked activity is higher in the daytime in green leaves of both bean plants and the duckweed *Lemna* — with reciprocal changes in the NAD-linked activity.[23,24] The light induced, reversible increase of the NADP enzyme in *Lemna* is inhibited by chloramphenicol and by amino acid analogs indicating that appearance of the enzyme is due to new protein synthesis possibly by chloroplast ribosomes.[25] Müller has found that both ATP and NADPH activate the NADP-linked glyceraldehyde phosphate dehydrogenase activity of spinach.[26] Thus two products of the light reactions might sustain the enzyme in a physical state which dictates the coenzyme specificity. ATP and NADPH do not affect the activity of NAD-linked glyceraldehyde phosphate dehydrogenase in this tissue. Thus the NAD-linked enzyme located outside the chloroplast cannot be activated by photosynthetic products which one would expect to remain inside the chloroplast in any case.

Nowhere is there decisive evidence to show that the two glyceraldehyde phosphate dehydrogenase activities in etiolated and green tissue are located either on separate proteins or on a single protein where the coenzyme specificity may be altered. Gibbs has made a heroic but as yet inconclusive effort to resolve this problem by enzyme purification, which is the only solid way to establish one of the two possible alternatives.[27] An NAD specific enzyme was purified from etiolated pea seedlings and an enzyme using either NAD or NADP was prepared from green pea tissue. There was no pronounced difference in K_M for substrate, product, coenzyme, or inhibitors. The two activities could not be separated on starch gel electrophoresis. Differences in stability and charcoal removal of bound coenzyme give a faint hint that these activities are due to different proteins, but this is hardly definitive proof. Yonuschot et al. have isolated what appears to be a pure glyceraldehyde phosphate dehydrogenase from spinach that uses both NAD and NADP at the same catalytic site.[28]

A similar situation exists among the algae: the NADP-linked activity is low in bleached *Euglena* and *Chlamydomonas* but rises as these cells turn green when shifted from heterotrophic to autotrophic growth.[29] A glyceraldehyde phosphate dehydrogenase has been purified from the blue-green alga *Anabaena variabilis* and it appears to use either NAD or NADP.[30]

Among the photosynthetic bacteria, *Chromatium* and *Rhodomicrobium vanneilli* have NAD specific glyceraldehyde phosphate dehydrogenases, which is reasonable since the bacteria seem to produce NADH as the reduced coenzyme of photosynthesis.[18] *Rhodopseudomonas palustris* may have an NADP-linked enzyme, since Kamen has presented presumptive evidence that this organism uses NADP instead of NAD as the coenzyme of photosynthesis.[31]

Fuller has discovered an interesting variant on the control of glyceraldehyde phosphate dehydrogenase activity in *Chromatium* as this organism adapts to the different growth conditions.[32] When the bacteria are grown in light on CO_2 with $Na_2S_2O_3$ as a reductant, the dehydrogenase has a lower K_M for substrate and product and a higher sulfhydryl content than the enzyme from the cells grown in light with malate as both the source of carbon and the source of reducing power. The K_Ms can be pushed up or down by chemical reduction or oxidation of the sulfhydryls in the isolated enzyme, suggesting that the growth environment can control the kinetic properties of the enzyme by altering the sulfhydryl content. The result is a higher substrate affinity when the enzyme must work hard fixing CO_2 for hexose synthesis and a lowering of that affinity when malate provides reduced carbon for cellular hexose.

In summary, glyceraldehyde phosphate dehydrogenase is NAD specific in etiolated higher plant tissue in which its only function is glycolysis or glucogenesis. An NADP-linked activity is activated by phytochrome to serve in photosynthetic metabolism and may even fluctuate on a day vs. night basis. In algae, the NADP vs. NAD activity shows a reciprocal variation depending on the photosynthetic competence of the cell. In bacteria, the activity is NAD-linked as is photosynthesis, but the K_Ms may vary in response to growth conditions.

2.5   THE PHOSPHOGLUCONATE OXIDATIVE PATHWAY

The phosphogluconate path is well-established in plant tissue both on the basis of isotope patterns and enzyme measurements. The overall pattern of reactions is described in Figure 2.2.

These reactions could provide little in terms of carbon skeletons — some C_4 for shikimic acid and a little C_5 for nucleosides but not the cell wall pentoses since these arise from decarboxylation of UDP glucuronic acid. The main function of this path apparently is to provide NADPH which could be important in an actively growing or differentiating tissue in which much biosynthesis demands NADPH but in which photosynthetic competence has not yet developed to meet this demand. The method of comparing the rates of appearance of isotope label in CO_2 from position 1 vs. position 6 of specifically labelled glucose has been used as a measure of the relative contributions of the phosphogluconate vs. the glycolytic pathway. Although this isotope method is subject to some artifact, it has proved reliable in certain applications to plant material. Gibbs and Beevers used this method to show a shift from glycolysis in embryonic tissue to the phosphogluconate pathway as the tissue matures.[33] Fowler and Ap Rees have applied the method to examine carbohydrate oxidation during differentiation of roots of pea seedlings.[34] By taking small segments along the root, they found that the undifferentiated cells at the root apex relied on glycolysis but the more differentiated tissue farther back showed an increase in phosphogluconate pathway relative to glycolysis. In the developing castor bean, one can measure a distinct contribution of reducing power from glucose via the phosphogluconate pathway to the massive fatty acid synthesis that is occurring in this oil-rich seed.[35] The phosphogluconate pathway seems to be the main route of carbohydrate oxidation in blue-green algae[36] which is perhaps reasonable in a creature with an incomplete Krebs cycle and a preponderance of NADP over NAD.

2.6   THE CITRIC ACID CYCLE

The citric acid cycle (Fig. 2.3) is operative in most plant tissue. As in heterotropic cells, this cycle must sustain the plant in darkness by conversion of

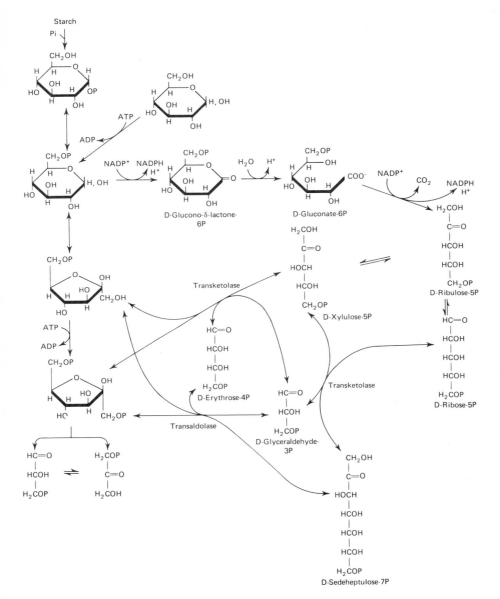

Figure 2.2   The phosphogluconate oxidative pathway or pentose phosphate shunt.

pyruvate to carbon dioxide, water, and energy. The substrates and enzymes have been demonstrated in many tissues and the respiration of most plant tissue can be inhibited by malonate, fluoroacetate, and arsenite (the latter blocking alpha

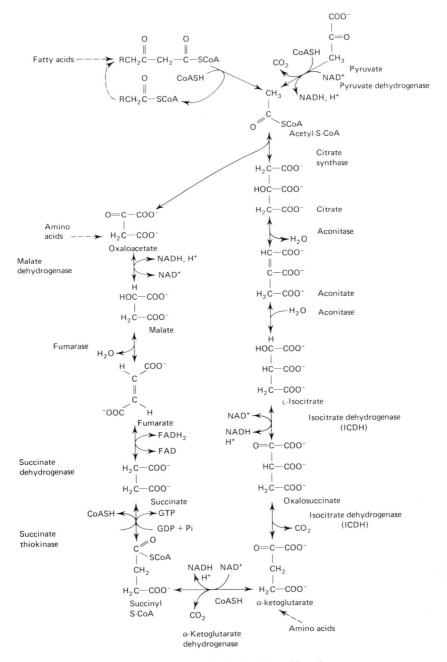

Figure 2.3   The Krebs citric acid cycle.

keto acid oxidation), all of which inhibit the Krebs cycle at the substrate level. As in other cells, this cycle is organized within the mitochondria. Most experiments have been done with tissue lacking chloroplasts, but there is sufficient evidence from green tissue to establish the Krebs cycle in both. Green leaves of higher plants are harder to work with since the mitochondria tend to have very high endogenous oxidase activity and tend to be contaminated with broken chloroplast fragments, but there is generally a standard Krebs cycle in a standard mitochondrion underneath all the difficulties.

Little has been done with the various enzymes of the cycle in plants other than to demonstrate their existence. Aconitase activity is often found to be diminished in iron deficient plants — tomato, soybean, etc. — which is suggestive evidence that iron is a prosthetic group on this enzyme but rarely if ever is the enzyme purified sufficiently to demonstrate iron as an integral part of the protein structure and there isn't a satisfactory iron removal-restoration to show iron involvement in the catalytic activity. Isocitric dehydrogenase activity shows interesting fluctuations in the cotyledons of germinating *Vignia sesquipedalis* — a bean.[37] The apoenzyme activity rises, then diminishes during germination but, more important, the virtual absence of NADP in this tissue seems to prevent Krebs cycle activity in favor of glyoxylate cycle activity. This tissue makes a good source of apo-isocitric dehydrogenase for NADP determinations. An NAD-linked isocitric dehydrogenase has been purified from etiolated peas[38] and could only be purified using 5 M glycerol to stabilize the activity. The enzyme activity appears to be controlled by the level of oxidized coenzyme present in that NAD activates and NADH inhibits catalysis.[39] Green plant material with its gross diurnal shifts from anabolism to catabolism might be the place to define the roles of NADP- and NAD-linked isocitric dehydrogenases. The ATP inhibition, AMP and ADP activation seen in the mammalian enzyme is not clearly established for the pea isocitric dehydrogenase. Isocitric dehydrogenase from peas is inhibited by oxaloacetate and glyoxylate — both are reasonable feedback regulators which if they accumulated in the cell could shut off further accumulation by stopping the isocitric dehydrogenase reaction.[40]

The alpha ketoglutarate dehydrogenase complex has been purified from cauliflower mitochondria and is similar to the enzyme complex isolated from mammalian sources.[41]

Many studies of malic dehydrogenase isozymes are extant and at least two mitochondrial and one cytoplasmic dehydrogenase have been demonstrated in corn root tissue.[42] In green tissue, yet another malic dehydrogenase is found associated with the chloroplast,[43] and this enzyme is NADP specific which is consistent with its location in a structure that generates NADPH in abundance.[44] One of the mitochondrial malic dehydrogenases is surely doing the heavy work of the Krebs cycle. Another malic dehydrogenase (cytoplasmic or mitochondrial?) might function in CO_2 fixation via reduction of PEP to malate. Two isozymes of malic dehydrogenase have been found in mitochondria from barley and these are the result of polymorphism of the mitochondria.[45] The plants used as the source of these mitochondria are hybrid and the nucleus contains genes for both of the malic dehydrogenases. The chloroplast malic dehydrogenase could function in a shuttle mechanism for moving either reducing power or organic acids across the chloroplast membrane. This shuttle mechanism is illustrated in Figure 2.4. The malic dehydrogenase in the chloroplast is activated by exposure of the chloroplast to light and in a subsequent dark period the enzyme loses activity.[46,47] Since treatment with the reducing agent dithiothreitol can achieve the same effect as illumination, one imagines that malic dehydrogenase is activated *in vivo* by a reduced product of photosynthesis.

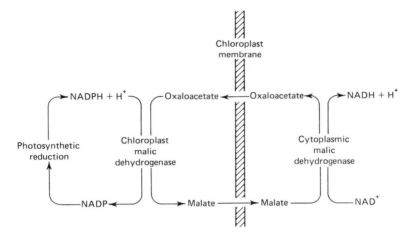

Figure 2.4   A shuttle mechanism for the export of reducing power across the chloroplast membrane.

Thus light, which generates reducing power within the chloroplast, could stimulate the export of that reducing power via a malic dehydrogenase shuttle into the cytoplasm. The cytoplasmic enzyme could function in the other half of this shuttle. Some of these isozymes are being sorted out from spinach leaves.[48] Still another malic dehydrogenase is found in glyoxysomes in which the glyoxylate cycle converts lipid via malate to carbohydrate.[49]

A special exception to the general occurrence of the Krebs cycle in plant tissue is emerging from studies of obligate autotrophs. Most higher plants and algae are capable of heterotrophic growth using reduced carbon supplied as a nutrient. Blue-green algae and some photosynthetic and chemosynthetic bacteria cannot grow in the dark when supplied with fixed carbon. It might be assumed that reduced carbon compounds do not penetrate the cell but this is disproved by isotope penetration measurements. There is growing evidence that the obligate autotrophs lack an effective alpha ketoglutarate dehydrogenase. Stanier

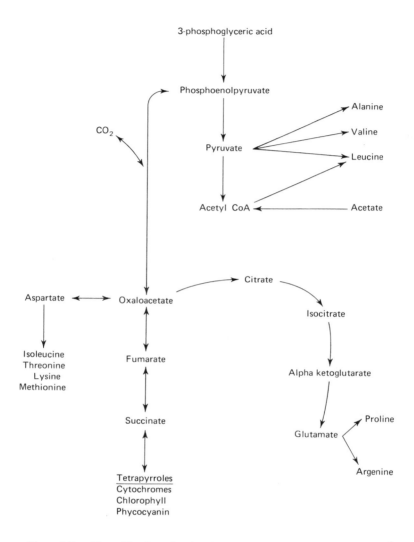

Figure 2.5 The utilization of carbon by obligate autotrophs. [From Smith, A. J., J. London, and R. Y. Stanier, *J. Bact.*, **94**, 972 (1967).]

has provided the most complete evidence by C^{14} feeding experiments and has detailed the isotope transfer through a host of reactions but found no movement of isotope from alpha ketoglutarate to succinate.[50] It was first suggested that these obligate autotrophs lack all respiratory chain activity and have no way to get rid of NADH, but this is probably incorrect. Later, evidence of respiratory activity in blue-green algae will be reviewed. Direct enzyme analysis in *Anabaena variabilis* and *Chloropseudomonas ethylicum* indicates that the alpha ketoglutarate dehydrogenase is missing.[51,52] The metabolism of phosphoglycerate in the blue-green algae and obligate chemoautotrophic thiobacilli can be described as in Figure 2.5. This partial metabolic map for the obligate autotrophs illustrates two of the main functions of the Krebs cycle. In non-green tissue or even green tissue in darkness, the complete cycle must run to meet the energy needs of the cell. In addition, the cycle must supply carbon skeletons for amino acids just as it does in the obligate autotrophs. The obligate autotrophs get their energy from either photosynthetic or chemosynthetic redox reactions not associated with carbon metabolism.

There are many indications that in green plants light may influence the rate of passage of carbon through the cycle. Marsh and Gibbs have provided excellent evidence for the turnover of all intermediates of the cycle in illuminated *Scenedesmus*, a green alga.[53] The rate of entry of acetate may be decreased by light-supported lipid synthesis, but the citric acid cycle continues to turn.

2.7   CARBOXYLATIONS AND DECARBOXYLATIONS

In higher plants, an essential feature of the Krebs cycle is the regeneration of oxaloacetate. If the cycle is to be used to supply carbon skeletons for amino acids and accumulated organic acids, this can hardly be done at the expense of shutting the cycle down. In addition to alpha ketoglutarate and oxaloacetate that may be withdrawn from the cycle via amination in order to supply glutamate and aspartate, citrate itself may be withdrawn. A citrate cleavage enzyme is available in the cytoplasm to provide acetyl CoA and oxaloacetate for synthetic process in that compartment. Citrate cleavage enzyme activity increases in the ripening Mango fruit presumably to facilitate the very active biosynthetic processes associated with ripening.[54] One must regenerate the dicarboxylic acid at least at the last stage so that on taking out a citrate or a ketoglutarate, an oxaloacetate can be put back to permit the next turn of the cycle. This is usually accomplished by one or another of the following reactions:

$$\text{NAD (P) H} + \text{pyruvate} + CO_2 \underset{\text{malic enzyme}}{\rightleftharpoons} \text{malate} + \text{NAD (P)}$$

$$\text{Phosphoenol pyruvate} + CO_2 \underset{\text{PEP carboxylase}}{\rightleftharpoons} \text{oxaloacetate}$$

$$\text{Pyruvate} + \text{ATP} + CO_2 + H_2O \underset{\text{pyruvate carboxykinase}}{\rightleftharpoons} \text{oxaloacetate} + \text{ADP} + \text{Pi}$$

These enzymes are widely available in plant tissue and permit the assumption that anything can be drained off from the Krebs cycle since these enzymes will replace the dicarboxylic acid. A few details on PEP carboxylase in plant tissue are known. In cotton leaves, there are three isozymes of PEP carboxylase.[55] Although these isozymes have not been localized to different subcellular compartments, one is bound to a particle. Multiple isozymes might serve in different compartments or might serve in different functions under different regulators in the same compartment. The PEP carboxylase from maize leaves is inhibited by physiological concentrations of oxaloacetic acid.[56] This seems to be a reasonable control mechanism since the oxaloacetate would stop its own accumulation under conditions in which there was insufficient NADPH or transamination partners to drain it away.

That plants accumulate intermediates of the Krebs cycle is known to anyone who has studied German. Lemon juice is 0.3 M citric acid, apple juice is 0.1 M malic, etc. In addition, the accumulations of oxalic, tartaric, and malonic acids by plants are well documented.

The accumulation of organic acids in the leaves of Crassulacean plants is the best studied case.[57] As much as 10 mg malate/gm fresh weight of leaf tissue accumulates at night and then disappears during the day. The plants fix CO_2 at night but hardly at all in the day. At night carbohydrate is converted to phosphoenolpyruvate that can be carboxylated via PEP carboxylase to oxaloacetate and is then reduced to malate.

If the plants are exposed to $C^{14}O_2$ in the dark, then the malate (which contains most of the isotope) is isolated and degraded, and the isotope is distributed as follows:

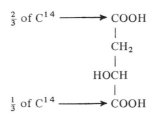

This is rationalized by assuming the pathway shown in Figure 2.6. The $\frac{1}{3}$, $\frac{2}{3}$ distribution of label persists in the dark for many hours without scrambling, indicating that the malate sits in a storage pool — possibly the vacuole. This malate storage pool is not limitless since, if the leaves are held in darkness too long, they begin to respire malate with an RQ = 1.33 (to oxaloacetate, to pyruvate, then around the cycle). When exposed to light, the Crassulacean leaves quickly deacidify and the carbon moves from malate to carbohydrate. In these circumstances, photosynthetic oxygen production exceeds CO_2 uptake from the atmosphere since malate decarboxylation provides the CO_2. These plants can thus afford the interesting habit of keeping their stomates closed in the daytime and thereby conserving water. Crassulacean plants can fix CO_2 at night by carboxylation and at that cooler time there is likely to be less water loss. There

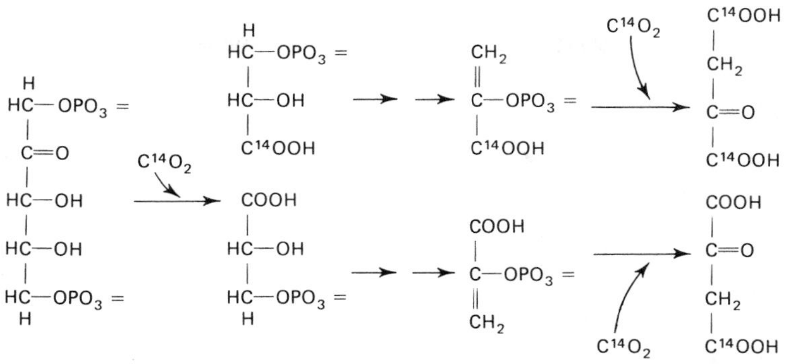

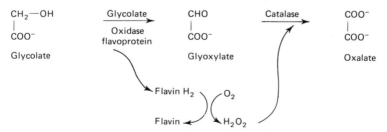

Figure 2.6   A suggested path explaining the distribution of isotope which occurs in Crassulacean acid accumulation.

is an effort to develop some species of *Crassulacea* as food crops in arid agricultural regions, e.g., parts of Mexico.

Malonate accumulates in a number of plant tissues. In non-legumes there is abundant acetyl CoA carboxylase (biotin containing).

$$\text{Acetyl CoA} + CO_2 + \text{ATP} \xrightarrow{Mg^{+2}} \text{malonyl CoA} + \text{ADP} + \text{Pi}$$

In beans, an oxaloacetic acid alpha decarboxylase has been characterized.

$$\tfrac{1}{2} O_2 + \text{oxaloacetate} \longrightarrow \text{malonate} + CO_2$$

Feeding C^{14} malonate shows its ready utilization by most plant tissues via acetyl CoA.

Several plants such as spinach, *Oxalis*, and a variegated ornamental called "dumb cane" accumulate oxalic acid. The oxalic acid probably arises from the oxidation of glycolate in the presence of catalase. Isotope labelled carbon dioxide is rapidly converted to oxylate (see Fig. 2.7) via glycolate and glyoxylate in *Oxalis corniculata.*[58]

CH$_2$—OH   Glycolate   CHO   Catalase   COO$^-$
| Oxidase | |
COO$^-$ flavoprotein COO$^-$ COO$^-$

Glycolate   Glyoxylate   Oxalate

Flavin H$_2$ O$_2$

Flavin   H$_2$O$_2$

Figure 2.7   The formation of oxalate by oxidation of glycolate in the presence of catalase.

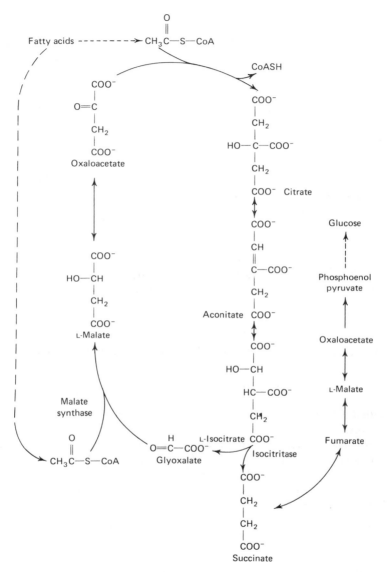

Figure 2.8   The glyoxylate cycle. These reactions convert acetate units from lipid via citrate, isocitrate, succinate to phosphoenolpyruvate which is used to synthesize glucose.

2.8   THE GLYOXYLATE CYCLE

In a number of plants such as castor bean and peanut, reserve carbohydrate in the seed is replaced by lipid. On germination this carbon reserve must be used to provide both substance and energy for the seedling until photosynthesis can begin. Prior to emerging from the soil, the seedling needs glucose to provide energy for growth and to provide building blocks for the cellulose and other hexose containing constituents of the cell. Hence, lipid must be converted to carbohydrate and this is done via the glyoxylate cycle. The reserve lipid is converted to acetyl CoA that condenses with glyoxylate in a unique reaction characteristic of this cycle to give malate. By reactions like those in the citric acid cycle the oxaloacetate and more acetyl CoA are converted to isocitrate. The isocitrate is then split in another unique reaction to succinate and glyoxylate. The glyoxylate produced is recycled and the succinate is exported from the cycle to be converted to hexose or oxidized to give ATP. The enzymes catalyzing this pathway, including a set of beta oxidation enzymes to convert fatty acids to acetyl CoA, are located in a microbody called the glyoxysome.[59,60] The glyoxysomes show a transient existence in the developing cotyledon (seed leaf) of the developing plant and disappear as the lipid reserves are exhausted and photosynthesis takes over.[61,62] In cotyledons of germinating cotton, the isocitritase activity, which is a key enzyme in the glyoxylate cycle, is uniquely depressed by chilling temperatures.[63] This cold inhibition would in turn retard growth until a more favorable temperature recurred.

Although the malate synthetase and isocitrate lyase are unique enzymes found only in the glyoxysome, the malic dehydrogenase, citrate synthase, and aconitase are found in both glyoxysomes and mitochondria. These later enzymes are probably distinct isozymes in each subcellular compartment. The citrate synthase in glyoxysomes and mitochondria of castor bean seedlings are distinct in their sensitivity to ATP inhibition.[64] As in most tissues, the mitochondrial enzyme is inhibited by ATP which is a rational feedback regulation in an ATP synthesizing organelle. The glyoxysomal citrate synthase is not inhibited by ATP and such regulation would not be expected in this organelle that has no obvious ATP synthesizing mechanism.

The catalysts within the glyoxysome accomplish the conversion of fatty acids to succinate. The succinate must move to the mitochondria where it can be converted to phosphoenol pyruvate, which can serve as a precursor to hexose via reverse glycolysis (Fig. 2.8).

GENERAL REFERENCES

Bonner, J., and J. E. Varner (eds.). *Plant Biochemistry.* New York: Academic Press, 1965.

Geissman, T. A., and D. H. G. Crout. *Organic Chemistry of Secondary Plant Metabolism.* San Francisco: Freeman, Cooper and Co., 1969.

Pridham, J. B., (ed.). *Plant Cell Organelles.* New York: Academic Press, 1968.

Robinson, T. *The Organic Constituents of Plants.* Minneapolis, Minn.: Burgess Publishing Co., 1964.

3 | ELECTRON TRANSPORT: THE REDUCTION OF OXYGEN, SULFATE, AND NITRATE

The respiratory chain associated with catabolic energy conservation from the Krebs cycle is generally similar in plants and animals. The sequence of electron carriers (Fig. 3.1) has been established by spectral and chemical identification of the individual participants, by determination of the redox potentials, kinetics of sequential reduction or oxidation of the carriers, isolation of partial reaction sequences, and by use of selective inhibitors. Since most of the measurements are spectroscopic, detailed studies have centered on non-green tissue so that the problem of chlorophyll masking is avoided. What is needed is a good fractionation method to separate mitochondria from chloroplasts and chloroplast fragments in homogenates of green tissue or application of more sensitive spectral methods developed for studies of photosynthesis to examine the mitochondria in green tissue. Still there is no serious reason to believe that the green and non-green tissue have different respiratory chains.

3.1 FLAVOPROTEINS, CoQ, CYTOCHROMES

As in all mitochondria, there are several flavoproteins that oxidize NADH.[1] All are sensitive to amytal. One is insensitive to rotenone (is this because the

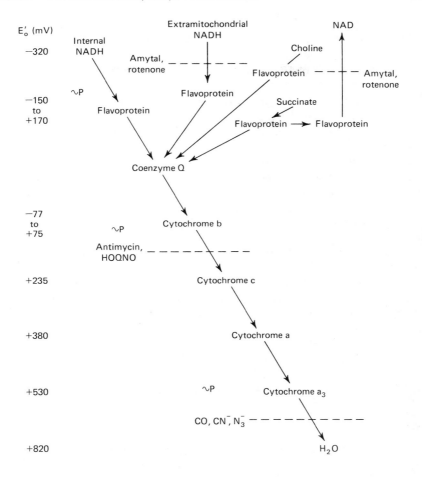

Figure 3.1   The respiratory electron transport chain in mitochondria.

rotenone binding site was altered in purification? Some electrons do get through from NADH in mitochondria treated with rotenone). A different flavoprotein may be uniquely involved in the reverse electron flow from FAD to NAD. In addition, there are specific flavoproteins that feed electrons into the co-enzyme Q region from succinate and from choline. There are several converging lines of evidence for a distinct flavoprotein on the outer membrane of mitochondrion which oxidizes external NADH.[2,3] Finding so many flavopro-teins in mitochondria only acknowledges their possible existence as independent components. It is not certain that some are multiple forms of the same enzyme created as artifacts of purification or inhibitor treatment. Storey has recognized five distinct flavoprotein components in a redox titration of flavoproteins in intact skunk cabbage mitochondria.[4] These flavoproteins ranged in potential from -155 to $+170$ millivolts and this method of recognition eliminates some of the potential artifacts in overestimating the number of flavoproteins in the mitochondrion.

Note that Figure 3.1 does not include non-heme iron proteins in the plant respiratory chain although it is extremely fashionable to include these proteins in mammalian and bacterial chains. Experimental evidence on this type of participant in the plant system is developing.

There has been a reluctance to put CoQ on the main chain of electron transfer since the kinetics of its redox changes appear to be slow compared to the cytochromes on changing the steady-state condition of the mitochondria. These measurements are made difficult by the large pool of CoQ relative to the amounts of cytochromes and by technical difficulties in observing CoQ absorbance changes. The CoQ extraction and restoration with concomitant loss and restoration of electron transport activity argue for a functional role of CoQ in the chain.

Evidence for the existence of three species of cytochrome b is spectroscopic and does not distinguish between sequential or parallel participation in the chain.[5] Difference spectra are done with the blank curvette containing aerated mitochondria where the cytochromes are fully oxidized and the optical density of this cuvette is then subtracted from that of the experimental cuvette that contains substrate or reducing agent to reduce cytochromes. Figure 3.2 illustrates difference spectra of potato mitochondria. Frequently something like cyanide or azide is added to prevent reoxidation of the reduced cytochrome. The absorption spectra are measured at liquid nitrogen temperature to reduce molecular motion and vibration and thus sharpen and intensify individual absorption bands. Cytochromes are characterized by three visible absorption bands called the alpha, beta, and gamma (or Soret) bands. The alpha band of an a type cytochrome appears at 590 nm or above. The b type cytochromes have alpha bands above 550, beta bands above 520, and the Soret or gamma band above 420 nm. Cytochromes of the c type have the peaks shifted slightly toward the blue — the alpha at 550, beta at 520, and the gamma at 420 or just below. In the spectra shown here, the alpha peak of both cytochrome a and cytochrome a_3 appears as a single peak at 598 nm, and this hump is eliminated by antimycin which prevents electrons from reaching this level of the respiratory chain. Cytochromes a and a_3 can be distinguished in other tissue by better separated alpha bands or by complexing a_3 with cyanide or azide.

The alpha and beta bands of cytochrome c appear at 547 and 516 nm and these drop out of the antimycin spectrum since cytochrome c reduction is blocked by antimycin. Thus, one can assign the 552, 557, and the 561 nm peaks as alpha bands of cytochrome b's above the antimycin block.

With succinate reduction under a nitrogen atmosphere, the cytochrome b molecules don't reduce as well — as though one is bypassed by the succinic dehydrogenase flavoprotein. In the succinate-nitrogen spectrum, the 561 nm

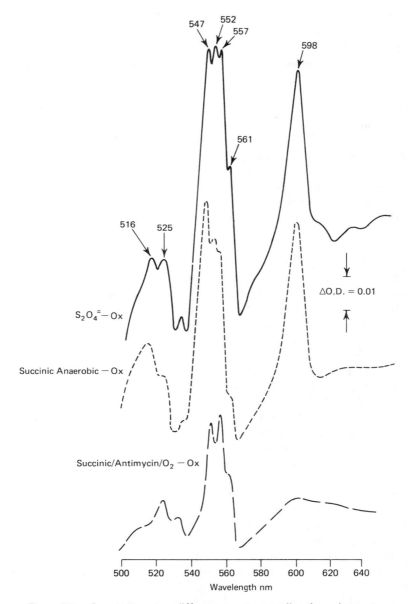

Figure 3.2   Low temperature difference spectra revealing the various cytochrome components in white potato mitochondria. [From Lance,
C. and W. D. Bonner, *Plant Physiol.*, **43**, 756 (1968).]

peak is very low and the 534 nm bump disappears, suggesting that a 561 alpha
and a 534 beta for that one of the b cytochromes that succinate does not reduce.

In the succinate plus antimycin spectra, the 527 beta band is best associated with the 552 and 557 alpha bands, suggesting that the 527 peak is a composite of two beta bands.

If you make a difference spectrum of dithionite reduced minus succinate-nitrogen reduced, the alpha band at 561 shows up clearly as the b type cytochrome reduced by dithionite but not by succinate. The same kinds of manipulations can be used in the Soret region to identify the gamma peaks.

The sequential arrangement of the electron carriers in the chain has been established by kinetic measurements and by selective inhibitor and selective electron donor experiments such as those described above. Dutton and Storey have recently measured the redox potentials of the cytochromes in plant mitochondria and, on the whole, their values are consistent with the sequence.

Using the molar extinction coefficient for the change in absorbance on reduction of purified cytochrome c (and there is evidence that other cyto-chromes have the same extinction coefficient as cytochrome c which is a convenient standard), one can calculate cytochrome concentrations in these mitochondria.

If the mitochondria are extracted with phosphate and cholate, cytochrome c comes out just as it does from rat liver mitochondria, but the 547 alpha and 515 beta bands remain in the mitochondria, indicating that a cytochrome c_1 is present buried under the cytochrome c 549 alpha and 517 beta bands in the unextracted mitochondria.

In terms of relative amounts, there is 1.00 cytochrome c + c_1, 0.65 cytochrome a + a_3, 0.67 cytochromes of the b type (about $\frac{1}{3}$ of each), 3.30 flavoprotein, and 14.1 NAD. This is about the ratio of carriers found in Jerusalem artichoke, mung bean, cauliflower, and skunk cabbage mitochondria. Note that there is no NADP in these plant mitochondria — unlike mammalian preparations. The Q_{O_2} of the mitochondria and the turnover of individual carrier cytochromes are quite similar among these plants and are similar to values for mammalian preparations. The Q_{O_2} (μliters/hr/unit tissue) of the tissues as opposed to the isolated mitochondria varies — 40 for artichoke, 4,000 for skunk cabbage — indicating that the number of mitochondria, not the rate of mitochondrial respiration, determines the respiratory activity of the cell.

Studies of oxidative phosphorylation with plant mitochondria give P/O ratios approaching the expected values and have yet to offer any special advantages in the problem of energy coupling. Uncoupler and inhibitor sensitivity of plant preparations is generally the same as rat liver mitochondria.[7,8] There is evidence for energy dependent reverse electron flow as seen in the reduction of NAD by succinate within mung bean mitochondria.[9] Even in the Arum spadix mitochondria, which appear to have lost most of their capacity for ATP synthesis in the oxidative direction, there is an indication that energy released by an exergonic redox reaction can drive endergonic reverse electron flow.[10]

3.2   SPECIAL PROBLEMS IN PLANT RESPIRATION

Cyanide Insensitive Respiration.

In terms of tissue respiration, there is a prominent characteristic of some higher plant materials with respect to inhibitor sensitivity. There is a cyanide and carbon monoxide insensitive respiration of *Aracea* — the skunk cabbage commends itself to the experimentalist not because of its foul odor but because the spadix of the emerging plant melts the snow around it, suggesting a high-respiratory activity with concomitant heat loss. Note from the preceding comparisons that the Q_{O_2} for skunk cabbage is one hundredfold greater than that for artichoke and that the mitochondria appear to be inherently uncoupled. In addition to the *Aracea*, sliced storage tissue of plants acquires a cyanide insensitive respiration, many green tissues are inhibited only 50%, and possibly all plant mitochondria can with age acquire some cyanide insensitivity. The suggestion that this cyanide insensitive respiration is the result of an auto-oxidizable b type cytochrome is weakly founded and the older suggestions that polyphenolase or ascorbic acid oxidase were the cyanide insensitive terminal oxidases have been discarded since these enzymes are not clearly associated with the mitochondria in which the cyanide insensitive respiration is located. The cyanide insensitive pathway apparently does not involve cytochromes a or a_3.[11] Cyanide insensitive respiratory activity is inhibited by hydroxamic acids[12] and by piericidin A and 2-thionyltrifluoroacetone[13] which latter compounds block the non-heme iron proteins in mammalian mitochondria.

Minimal Respiration of Autotrophs.

The blue-green algae are another unique case since these obligate photoautotrophs were assumed to have no need for a respiratory chain and since respiration was at best very hard to measure in these cells. There is clear evidence for oxygen consumption by these algae, although the rate is low and evidence for inhibition of this respiration by light, or more accurately inhibition by photosynthesis, has been presented.[14] This inhibition of respiration by photosynthesis suggests a competition between the two processes for a nucleotide coenzyme. Using cell-free preparations, Smith et al. could find no NADH oxidase[15] but Leach and Carr reported oxidase activity using reduced pyridine nucleotides as electron donors.[16] The NADPH rate calculates to 0.27 μl O_2/mg protein/hour which is low compared to the rate of 50 μl O_2/mg protein/hour for artichoke mitochondria. This figure for the algae would be helped a bit if the oxidase activity were purified to the same extent as mitochondria are purified vis-à-vis crude cell extract. Horton has found an NADH oxidase activity in blue-green algae extracts that appears to be bound to large particles and this activity is cyanide insensitive, azide insensitive, and about

50% inhibited by HOQNO, amytal, and rotenone.[17] Leach and Carr have reported low but plausible rates of oxidative phosphorylation with blue-green algae preparations.[18] Hackett approached the problem of blue-green algae respiration by studying colorless organisms, e.g., *Vitreoscilla*, which are presumed to be albino forms of blue-greens.[19,20] His spectroscopic studies indicated several heme proteins, one of which appeared similar to cytochrome o that is thought to be the terminal respiratory carrier in bacterial respiratory chains.

Salt Respiration.

Salt respiration is a problem of long standing with plant materials in that roots actively absorb salt against a concentration gradient so there must be energy input from the respiratory chain. Many root tissues, when immersed in salt solution, show an increase in respiratory activity. Lundegard had proposed that anions might move up the respiratory chain as electrons moved down and the stoichiometry of ion uptake to electron movement is not inconsistent with this proposal. Dinitrophenol blocks ion uptake but stimulates respiration and this led to the contemporary view that the mitochondria might act as ion pumps using energy otherwise conserved by phosphorylation. Current evidence suggests that ion transport in root tissue is directly coupled to electron transport and may be an alternative to ATP formation rather than requiring ATP *per se.*[21]

In animal tissue, the ion transport ATPase is now a prominent object of study. An ATPase stimulated by sodium and potassium ions is found in membranes and is inhibited by ouabain that blocks active transport of sodium and potassium ions. The enzyme is presumed to couple the energy released by ATP cleavage in order to pump the ions across a membrane. That such a mechanism might occur in plant tissue — distinct from and more general than the respiratory linked root-pumping system — has been suggested but awaits experimental proof.

Wound Respiration.

Wound respiration refers to the rise in respiratory activity in plant tissue that is observed when a potato is sliced. The rise is not instantaneous and since it can be prevented by actinomycin D and puromycin, it appears to depend on RNA and protein synthesis.[22,23] One can measure increases in succinic dehydrogenase, cytochrome oxidase, mitochondrial nitrogen, and phospholipid indicating that the number of mitochondria increases. In addition, a new isozyme of polyphenolase appears in cut and aerated potato slices.[24] Concurrent with these increases, artichoke tuber shows a transient rise in the ability of cell-free preparations to incorporate C^{14} labelled leucine.[25] One has the impression that a large area of the genome is derepressed in order to increase the respiratory activity (the new increment is frequently cyanide insensitive and possibly distinct from the pre-induced respiratory chain) and other enzymes as well (invertase, phenylalanineammonia lyase).

Ripening Climacteric.

A climacteric rise in respiration accompanies ripening in certain fruits in which a peak in respiration is achieved when fruit development and maturation are complete and this is followed by a respiratory drop off as the tissue senesces. Although many attempts to relate this transient to substrate, phosphate, and/or ADP availability or uncoupling of the mitochondria have failed, there is evidence of increased mitochondrial synthesis concurrent with climacteric onset.

3.3   GENESIS AND PARTIAL AUTONOMY OF MITOCHONDRIA

The synthesis of mitochondria in germinating seeds offers interesting possibilities for study of organelle assembly. There is a rapid increase in the activities and concentrations of mitochondrial components during the early stages of germination of the pea.[26] In the peanut, the dry embryo respires very slowly and yields mitochondria deficient in cytochrome c and lacking any respiratory control.[27] When the embryo becomes hydrated by imbibition, cytochrome c is synthesized and the respiration becomes controlled to allow coupling of the energy released by oxidation to ATP synthesis.

The biogenesis of mitochondria and the extent of their autonomy from the nucleus is now an area of considerable interest. The fact that mitochondria have their own complement of DNA suggests that at least part of the mitochondrial protein is coded for outside the nucleus. Several careful isolations have been done on plant mitochondrial DNA.[28,29] There is a recent report on DNA isolated from pea leaf mitochondria under very gentle conditions. More than half of the isolated DNA is in the form of circles 30 μ in diameter. Several methods of measurement indicate a molecular weight value of 70×10^6 and the renaturation kinetics of the sheared DNA indicates a single kinetic class of molecules and gives no suggestion of repeat sequences.

If 340 is used as the molecular weight of an average nucleoside monophosphate unit, the total number of nucleotides per molecule of DNA is calculable.

$$\frac{70 \times 10^6}{340} = 2 \times 10^5 \text{ nucleotides}$$

One can assume half of these nucleotides are involved in protein coding and 3 nucleotides are needed per amino acid coded.

$$\frac{1 \times 10^5}{3} = 33,000 \text{ codons}$$

If one assumes, gratuitously, that the average protein in mitochondria has 200 amino acids (molecular weight approximately 24,000), the number of proteins coded by mitochondrial DNA can be estimated.

$$\frac{33,000}{200} = 165$$

proteins could be coded on this mitochondrial DNA. This calculation neglects the requirement for DNA in coding ribosomal and transfer RNAs for the mitochondria, but this is not likely to be a large percentage of the total informational capacity. Mammalian mitochondrial DNA is known to be circular with a contour length of 4.5 to 6 microns. The plant mitochondrial DNA is eight to ten times the length of the animal mitochondrial DNA suggesting that it either contains ten times as much information or is much more repetitious. Since redundancy in the polyploid chromosomes of the plant nucleus is common, one suspects that plant mitochondrial DNA may be redundant despite the evidence from renaturation kinetics. Soon plant mitochondrial rRNA, mRNA, and tRNA will be characterized as they have been for yeast. The yeast mitochondrial ribosomes have 50S and 30S pieces to give a 70S ribosome like those of bacteria and unlike the yeast cytoplasmic ribosome which is an 80S particle formed from 60S and 40S pieces. Note that cycloheximide inhibits 80S but not 70S ribosome function, but chloramphenicol inhibits 70S but not 80S ribosomes. One must wait for extensive hybridization studies in order to clarify the interrelations of DNA and RNA from the mitochondria with the nucleic acids of the nucleus, chloroplast, and the cytoplasm. By analogy with yeast and mammals, it is unlikely that plant mitochondria are entirely autonomous or that their DNA can code for more than a small fraction of the mitochondrial protein. In mammalian mitochondria, it seems that the mitochondrial DNA codes for structural protein, a few coupling factors, and the mitochondrial ribosomes, but the outer membrane proteins, surely cytochrome c and probably the soluble dehydrogenases, are coded in the nucleus and synthesized on cytoplasmic ribosomes.[30,31]

3.4   SULFATE REDUCTION

Sulfate reduction is a major metabolic phenomenon by which plants provide both themselves and most heterotrophs with reduced sulfur for amino acids.[32] Mammals are incapable of reducing sulfate. Inorganic sulfate is 8 electrons deficient compared to the reduced sulfur in protein. One might assume that addition of pairs of electrons in a stepwise fashion would be the most likely sequence of reduction, but the chemistry of sulfur does not supply enough encouraging intermediates.

$$\text{Valence } +6 \qquad +4 \qquad +2 \qquad 0 \qquad -2$$

$$SO_4^= \longrightarrow SO_3^= \longrightarrow X \longrightarrow X' \longrightarrow S^= + 4\,H_2O$$

An enzymatic pathway that covers at least part of the ground is known and the cell-free reduction of $SO_4^=$ to $S^=$ can be observed. It is first necessary to activate sulfate and the enzyme that does this, ATP sulfurylase, has been purified from yeast and recognized in higher plants.

$$ATP + SO_4^= \longrightarrow AMP - SO_4^= + PP_i$$
$$\downarrow$$
$$2\,P_i$$

The isolated ATP sulfurylase is not specific for $SO_4^=$, but it will react with any group IV anion — chromate, molybdate, or tungstate. These reactions turn the enzyme into an ATP pyrophosphorylase, presumably via the following sequence:

$$ATP + enzyme \longrightarrow AMP\ enzyme + PP_i$$

$$
\begin{array}{ccc}
 & MoO_4^= & & MoO_4^= \\
AMP - enzyme + & CrO_4^= & \longrightarrow AMP - & CrO_4^= + enzyme \\
 & WO_4^= & \swarrow & WO_4^= \\
 & & AMP &
\end{array}
$$

Only a catalytic amount of any of these anions is needed to support the phosphorolysis of ATP. This phenomenon may explain the toxicity of these anions and such reactions indicate that the sulfurylase is operating through an AMP intermediate.

It is worth noting that selenate ($SeO_4^=$) can be activated by the same ATP sulfurylase. Some plants can reduce the selenate and incorporate it into selenomethionine. Since this amino acid analog is toxic, it constitutes a serious problem in the Dakotas where the plants of the genus *Astragulus* concentrate selenium from the soil to give levels of selenomethionine fatal to grazing animals.

The product of the ATP sulfurylase, AMP - SO_4, called APS or adenosine phosphosulfate, must be further activated by phosphorylation with ATP.

$$APS + ATP \longrightarrow PAPS + ADP$$

The enzyme that phosphorylates APS gives $3'$phosphoadenosine $5'$phosphosulfate and is called APS kinase. This enzyme has been found in several species in the chloroplast where it is conveniently close to the source of ATP generated by photosynthetic phosphorylation.[33,34] This reaction, in addition to the pyrophosphatase pulling the sulfurylase step that precedes it, gives a large negative free-energy change to help the strongly endergonic sulfurylase reaction.

The $SO_4^=$ is now sufficiently prepared for reduction and the reduction process is beginning to yield information of a substantial sort. Both higher plants and yeast show the same reactions:

$$NADPH + PAPS \longrightarrow NADP + PAP + SO_3^=$$

Further reduction of sulfite to sulfide is catalysed by a complex group of proteins including an enzyme A, an enzyme B and a protein fraction C. Enzyme A is a flavoprotein with diaphorase activity and is inhibited by reagents that block vicinal dithiols. The flavoprotein apparently acts by reducing a disulfide bond in fraction C.

$$\text{fraction C} \underset{\diagdown}{\overset{\diagup}{\underset{S}{\overset{S}{\big|}}}} + NADPH \xrightarrow{\text{enzyme A}} \text{fraction C} \underset{\diagdown}{\overset{\diagup}{\underset{SH}{\overset{SH}{\big|}}}} + NADP$$

If $S^{35}O_4^=$ is incubated with the enzyme system, one finds labelled product $S^{35}O_3^=$ attached to fraction C through a sulfhydryl and this bound sulfite is further reduced to sulfide:

$$PAPS + \text{reduced fraction C} \longrightarrow \text{fraction C} \underset{\diagdown}{\overset{\diagup}{\underset{SH}{\overset{S - SO_3^-}{}}}} + P_i + AMP$$

$$\text{fraction C} \underset{\diagdown}{\overset{\diagup}{\underset{SH}{\overset{S - SO_3^-}{}}}} + 6 \text{ electrons} \xrightarrow{\text{enzyme B}} \text{fraction C} \underset{\diagdown}{\overset{\diagup}{\underset{S}{\overset{S}{}}}} + S^=$$

Enzyme B is a sulfite reductase and has been studied using reduced methyl viologen dye as an electron donor. A partially purified sulfite reductase has been obtained from spinach[35] and appears to be localized in the chloroplast.[36] This enzyme has been purified 500 fold and is recognizable as a heme protein.[37,38] The molecular weight is estimated at 85,000. NADPH cannot reduce this heme protein but reduced ferredoxin does reduce it readily and is the likely donor coupling sulfite reduction to photosynthesis. Since the preparation is inhibited by carbon monoxide and this inhibition is light reversible, in this respect the heme protein resembles an a type cytochrome, but the redox potential must be amazingly low for any cytochrome. Schmidt and Trebst have elaborated a similar series of reactions in spinach chloroplasts and suggest that glutathione reductase might be a participant in the reduction of sulfate to sulfite.[39]

One might object that the six electrons needed to convert sulfite to sulfide seems rather much for one step, especially if these electrons come via a cytochrome that is probably a one-electron carrier. Nevertheless, no intermediates between $SO_3^=$ and $S^=$ have been found and there are no bacterial mutants blocked at this point in the bacterial sulfite reduction pathway.

The reduced sulfide is evidently introduced into the organic sulfur pool mainly via the serine sulfhydrase reaction.

$$
\begin{array}{ccc}
\hspace{3.5cm}CH_2\,OH & \hspace{3cm}CH_2\,SH \\
\hspace{3.5cm}| & \hspace{3cm}| \\
H_2 S + NH_2 - CH - COOH \rightleftharpoons H_2 O + NH_2 - CH - COOH
\end{array}
$$

The metabolism of sulfate in green algae appears to be quite similar to that of higher plants.[40]

3.5   NITRATE REDUCTION

Nitrate reduction is another major reductive process in plants, judging simply from the enormous quantity of nitrate that is spread on crop land.

Again, the process involves the introduction of eight reducing equivalents, which in conventional terms might be supposed to proceed by addition of two electrons at different steps.

$$
\begin{array}{cccccc}
\text{Valence} & +5 & +3 & +1 & -1 & -3 \\
\\
& NO_3^- \longrightarrow & NO_2^- \longrightarrow & X \longrightarrow & X' \longrightarrow & NH_3
\end{array}
$$

A number of compounds like nitrous oxide, nitroxyl or its dimer, nitramide, and hydroxylamine might serve as intermediates but all are exceedingly unstable, toxic, and might only function as protected enzyme bound intermediates.

The reduction of nitrate to nitrite is a well-documented reaction in plant material and a nitrate reductase has been purified from a number of species. Nitrate reductase is a molybdoflavoprotein. The Neurospora enzyme is NADPH specific while the enzyme from higher plants is often NADH specific. The higher plant enzyme apparently resides in the cytoplasm and carbohydrate exported from the chloroplast provides the reducing power as NADH.[41] Nitrite and ADP act as inhibitors of spinach nitrate reductase and both would be reasonable feedback signals since an excess of either nitrite or ADP would mean that NADH was needed elsewhere.[42]

Studies of the synthesis of nitrate reductase have yielded several very interesting observations. When *Chlorella* or spinach are grown on molybdenum deficient medium, the apoenzyme appears in the cells and is able to catalyze the transfer of electrons from NADH to an artificial dye. Molybdenum apparently is inserted onto the previously synthesized protein to make the complete nitrate reductase holoenzyme.[43,44] Although neither nitrate nor molybdenum is required for nitrate reductase synthesis, ammonia will repress the synthesis of this enzyme. In *Chlorella*, ammonia both represses and inactivates the nitrate reductase system.[45]

Nitrate reductase in radish is clearly an inducible enzyme.[46] Nitrate as sole nitrogen source brings the activity up and nitrite represses the nitrate induced activity. Nitrate reductase induction is inhibited by actinomycin D, puromycin, and cycloheximide indicating that the rise in enzyme activity is caused by *de novo* protein synthesis. Cultured tobacco cells provide a rather elegant example of induction of nitrate reductase and here the enzyme is repressed by casein hydrolysate.[47,48] In the intact plant, induction of nitrate reductase appeared to require light presumably for products of photosynthesis, but suitable concentrations of kinetin and gibberellic acid can eliminate the light requirement.[49] Corn seedlings will lose their ability to maintain polyribosomes and to form nitrate reductase on prolonged deprivation of light.[50] Exposure to light can allow reassembly of the polysomes and reactivates inducible nitrate reductase synthesis.

Nitrite reductase, on purification from higher plants and algae, appears to use ferredoxin as its electron donor.[51,52,53,54,55]

$$NADPH \longrightarrow \begin{array}{c} \text{chloroplast} \\ \text{NADP-ferredoxin} \\ \text{oxidoreductase} \end{array} \longrightarrow \text{ferredoxin} \longrightarrow \begin{array}{c} \text{nitrite} \\ \text{reductase} \end{array} \longrightarrow \begin{array}{c} NO_2 \\ \diagdown \\ NH_3 \end{array}$$

Various claims about the nitrite reductase as a flavoprotein need more investigation. Zumft has purified a nitrite reductase from *Chlorella* and this enzyme appears to be a non-heme iron protein.[56] This highly purified nitrite reductase preparation will also reduce hydroxylamine, although at a slower rate than nitrite. In higher plant tissues, hydroxylamine reductase can be fractionated away from nitrite reductase, although both activities give ammonia as the reduction product.[57]

Once NH_3 is produced, it enters the amino acid pool via glutamic dehydrogenase.

$$NH_3 + \text{alpha ketoglutarate} + NADPH \text{ (or NADH)} \rightleftharpoons$$
$$\text{glutamate} + H_2O + NADP \text{ (or NAD)}$$

The glutamic dehydrogenase from mung bean shows a fourfold better activity with NADH and this enzyme is loosely associated with the mitochondrion. An NADPH specific glutamic dehydrogenase has been found in the chloroplast which might be a more opportune site of amination since the nitrite to ammonia conversion appears to be localized there in addition to the most generous source of reduced pyridine nucleotide in the cell.[58]

An important manipulation of nitrogen metabolism may be available from studies with the herbicide Simazine − 2-chloro-4, 6-bis ethylamino 5 triazine.[59] When sprayed on rye, Simazine causes a 79% increase in protein. Since acrylamide gel electrophoresis of extracted protein shows a normal pattern, the Simazine does not cause excess synthesis of any one protein but rather a rise in all types. There is a marked increase in nitrate reductase activity in nitrate grown plants to serve the Simazine induced protein synthesis. What control point is affected by the herbicide is unknown, but the economic implications of this discovery are enormous.

GENERAL REFERENCES

Beevers, L., and R. H. Hageman. "Nitrate Reduction in Higher Plants," *Ann. Rev. Plant Physiol.*, **20**, 495 (1969).

Ikuma, H. "Electron Transport in Plant Respiration," *Ann. Rev. Plant Physiol.*, **23**, 419 (1972).

Tewari, K. K. "Genetic Autonomy and Extranuclear Organelles," *Ann. Rev. Plant Physiol.*, **22**, 141 (1971).

4 ‖ PHOTOSYNTHETIC CARBON METABOLISM

All of the carbon atoms in a green plant must ultimately arrive by a light driven fixation of atmospheric carbon dioxide. Although heterotrophic tissues can fix minor amounts of carbon dioxide, the relatively enormous amounts of fixation in autotrophic cells is accomplished using distinctive enzymatic pathways. The light generated driving force for these fixation pathways in plants can be conveniently postponed (Chap. 7, 8, 9) in order to concentrate attention on the metabolism of carbon *per se*. Although the general outline of most carbon fixation became known shortly after the introduction of radioisotopes, new insights continually draw attention back to this problem.

4.1   THE CALVIN CYCLE

Only a brief review of the Calvin cycle is needed here since the basic transformations are familiar in kind if not in detail from general surveys of carbohydrate metabolism. This major path of CO_2 fixation was elucidated by

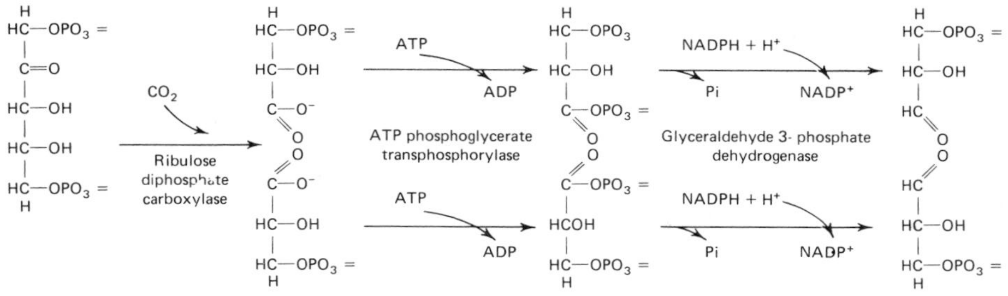

Figure 4.1   The initial reaction sequence for photosynthetic carbon dioxide reduction. Note that two moles of glyceraldehyde-3-phosphate can be converted to hexose without further expenditure of NADPH + H$^+$ or ATP.

using $C^{14}O_2$ and paper chromatography to isolate and identify intermediates. The kinetics of passage of isotope through the intermediates and even passage of isotope pulse through specific positions within the intermediates established the sequence of reactions. Ribulose diphosphate is carboxylated to yield two moles of phosphoglyceric acid which are then reduced to the level of the bulk carbohydrate of the cell. This reductive sequence (Fig. 4.1) requires ATP and NADPH generated by photo acts in amounts proportional to the amount of carbon dioxide fixed. Most of the reactions of the photosynthetic carbon cycle are arranged to regenerate ribulose diphosphate for the next round of carbon dioxide fixation. The eight participating reactions of the regenerative cycle are as follows (Fig. 4.2):

1. Conversion of fructose 1,6 diphosphate to fructose 6 phosphate by a **phosphatase**.

2. Transfer of the top two carbons of fructose 6 phosphate to glyceraldehyde 3 phosphate by **transketolase** to give erythrose 4 phosphate and xylulose 5 phosphate.

3. Condensation of erythrose 4 phosphate and dihydroxyacetone phosphate to give sedoheptulose 1,7 diphosphate by **aldolase**.

4. Sedoheptulose 1,7 diphosphate to sedoheptulose 7 phosphate by a **phosphatase**.

5. Transfer of the top two carbons of sedoheptulose 7 phosphate to glyceraldehyde 3 phosphate by **transketolase** to give **ribose 5 phosphate** and **xylulose 5 phosphate**.

At this stage there are three moles of **pentose** to convert to ribulose 1,5 diphosphate.

6. Ribose 5 phosphate is isomerized to ribulose 5 phosphate by phosphopentose **isomerase**.

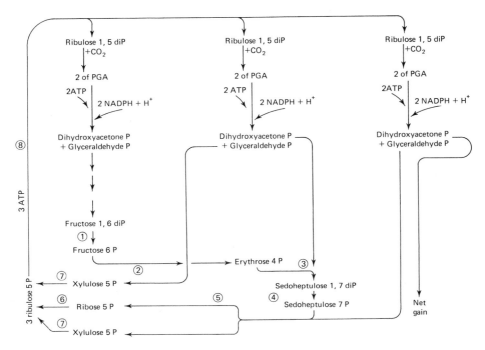

Figure 4.2   The Calvin cycle for photosynthetic fixation of carbon dioxide.
Note that six ATP and six NADPH are used to fix $3CO_2$. Three
more ATP are used to regenerate the CO_2 acceptor. The
numbered reactions refer to the regenerative cycle in the text.

7. Xylulose 5 phosphate is converted to ribulose 5 phosphate by
phosphoketopentose **epimerase**.

8. ATP and ribulose 5 phosphate give ribulose 1,5 diphosphate via
phosphopentose **kinase**.

4.2   RATES AND CONTROL PROCESSES

Several experimental observations lead to interpretive difficulties with the Calvin
cycle, but these difficulties are probably not serious. There is an asymmetric
distribution of C^{14} into the C_3 and C_4 positions of hexose. Although the cycle
as written would predict symmetrical distribution of isotope in both halves of
the hexose, there is no clear experimental explanation for this asymmetry. The
ribulose diphosphate carboxylase seems to have too low an affinity for carbon
dioxide in *in vitro* measurements to account for the *in vivo* activity. The K_M of
the isolated enzyme is unreasonably high for the *in vivo* condition. This K_M is
reduced in the best chloroplast preparations which fix CO_2 at a rate equal to
that of the intact leaf. When the enzyme is extracted from these chloroplasts,

however, its affinity for substrate is decreased. Possibly the enzyme is damaged on liberation from the chloroplast or the chloroplast holds the enzyme in a special configuration for efficient catalysis or the chloroplast may contain allosteric regulators. Cooper has shown that carbon dioxide, not bicarbonate, is the substrate for the ribulose diphosphate carboxylase.[1] Although the *in vivo* concentration of carbon dioxide appears to be low, the carbonic anhydrase present in chloroplasts may assist in concentrating the substrate.[2]

The Calvin cycle for CO_2 fixation was elucidated through isotope studies with intact plant cells. Next came a demonstration of all of the Calvin cycle enzymes in green leaf tissue and these enzymes were shown to be in the chloroplast. It appeared that the amount of activity for each step might not be sufficient to give a rate of CO_2 fixation commensurate with the rate of the intact plant. A mixture of purified enzymes plus a catalytic amount of ribulose diphosphate could do net CO_2 fixation in the dark if supplied with sufficient ATP and NADPH. With the discovery that illuminated chloroplasts could generate ATP and NADPH, it seemed that chloroplasts should be able to do complete photosynthesis.

The problem of chloroplast CO_2 fixation is discussed very nicely by Gibbs in a chapter called "The Rate Race."[3] The intact leaf can fix CO_2 at rates of 180 to 200 μmoles/mg chlorophyll/hr. The early chloroplast experiments showed fixation rates of 1 to 5 μmoles CO_2/mg chlorophyll/hr. Accepting the stoichiometry of 3 ATP and 2 NADPH per CO_2 fixed, one would require a rate of 600 μmoles ATP and 400 μmoles of NADPH/mg chlorophyll/hr produced by the chloroplasts. Such rates are routinely observed in measurements of photophosphorylation and Hill reaction activities of isolated chloroplasts. By refining the method of chloroplast isolation, Walker[4] and Jensen[5] pushed the rates of CO_2 fixation up to that of the *in vivo* reaction. The main trick is a very rapid separation of the chloroplasts from the rest of the homogenate. Jensen put leaves in a special cloth bag, squeezed out the juice into buffer, and centrifuged out the chloroplasts immediately to reduce the preparation time to two minutes. These chloroplasts are distinctive not only in their high CO_2 fixation rate but also in their appearance under light or phase contrast microscopy. The chloroplasts appear to be surrounded by a halo or membrane that is absent from chloroplasts isolated by more leisurely methods. Although the amount of protein per mg chloroplasts is increased fivefold by retention of the chloroplast membrane, the specific activity of individual enzymes of the Calvin cycle is not enormously increased (some may double or triple but none go up 100–200 fold as does the CO_2 fixation rate). The major advantage seems to be in keeping the enzymes properly organized within the chloroplast membrane.

With an intact chloroplast system, Walker noted a lag in the onset of photosynthetic CO_2 fixation, similar to the lag in the intact plant.[6] Walker supposed this lag might be due to loss of some Calvin cycle intermediates and found that inorganic phosphate, ribose-5 phosphate and phosphoglyceric acid would overcome this lag.[7] Inorganic phosphate alone, although alleviating the lag in onset of CO_2 fixation, reduces the maximal rate of CO_2 fixation[8] and this inhibition by phosphate is reversed by adding phosphoglyceric acid but not by ribulose diphosphate. Although inorganic phosphate inhibits the isolated ribulose diphosphate carboxylase, this inhibition is not reversed by phosphoglyceric acid; thus the ribulose diphosphate carboxylase is not the site of phosphate-phosphoglycerate regulation in the chloroplast system. It appears that phosphoglycerate, by allowing oxygen evolution via its reduction with NADPH,[9] is somehow regulatory — perhaps by allowing stoichiometric phosphorylation which accompanies NADP reduction by chloroplasts.

Bassham[10] has approached the problem of regulation of chloroplast CO_2 fixation with alternating light-dark exposures to $C^{14}O_2$ and his data suggest a possible light activation of the ribulose diphosphate carboxylase activity. By studying the kinetics of labelled intermediates coming out of the chloroplast, Bassham found distinct classes of intermediates that are either lost or retained by this organelle. Ribulose diphosphate, sedoheptulose phosphates, and hexose phosphates are retained by the chloroplast. Phosphoglyceric acid, dihydroxyacetone phosphate, fructose diphosphate, and pentose phosphates diffuse out of the chloroplasts, perhaps to support biosynthetic needs in other parts of the cell. There is even a suggestion that the selective diffusion of these metabolites through the outer chloroplast membrane might be controlled by fructose diphosphate phosphatase.[11] If one notes that the fructose diphosphate phosphatase seems to be used twice — to split off the phosphate at carbon number one from both fructose diphosphate and sedoheptulose diphosphate — this enzyme is a choice site for regulation. Fructose diphosphate phosphatase in the chloroplast may be controlled by photosynthetic activity.[12] Reduced ferredoxin, which might increase in steady-state concentration under conditions of intense photosynthetic activity, was found to activate up to tenfold the fructose diphosphate phosphatase in the chloroplasts. This activation is mediated by a protein that presumably catalyzes a reduction of the phosphatase by reduced ferredoxin. This fructose diphosphate phosphatase is not a conventional magnesium requiring enzyme with an alkaline pH optimum and is probably a uniquely controlled enzyme specifically operating in photosynthetic carbon metabolism. The ribulose phosphate kinase is also activated by illuminated chloroplasts or in the dark by chloroplast enzymes and dithiothreitol, again suggesting that activation is by way of a light generated reducing agent. Reference 13 contains a humorous view of experimental analyses of the Calvin cycle.

4.3   RIBULOSE DIPHOSPHATE CARBOXYLASE

Ribulose diphosphate carboxylase has attracted considerable attention as an enzyme unique to the Calvin cycle.[14] Its role in photosynthesis was assured by

Levine's discovery of a *Chlamydomonas* mutant lacking this enzyme and unable to fix CO_2.[15] Lane, succeeding when many others had failed, purified the enzyme from spinach and got a homogeneous protein of molecular weight 557,000.[16] Lane then split the spinach enzyme into subunits with sodium dodecyl sulfate and recognized two types of subunits on gel electrophoresis and in the ultra centrifuge.[17] The two subunits have different amino acid composition and show molecular weights of 55,800 and 12,000.[18] There are eight large subunits that contain catalytic sites and eight small structural or regulatory subunits in the spinach ribulose diphosphate carboxylase. This enzyme has now been isolated and characterized from a variety of sources and among algae and higher plants, the molecular weight and subunit composition seems quite uniform.[19,20] There are indications of greater variability among species in the amino acid sequence of the small subunit than in the large subunit.[19,21]

Kawashima has found non-synchronous incorporation of labelled carbon dioxide into the two subunits of this protein.[22] A pulse of photosynthetically incorporated isotope puts more label in the amino acids of the large subunit than those of the small one. This result suggests either a large pool of preformed small subunits or two ribosomal origins under different forms of regulation. The latter interpretation is supported by the work of Andersen et al., who studied mutations expressed in ribulose diphosphate carboxylase in various strains of tomato.[23] Different genetic strains representing several linkage group markers showed different alterations in the carboxylase. Smillie had reported the inhibition of ribulose diphosphate carboxylase synthesis by chloramphenicol indicating that chloroplast ribosomes are the site of synthesis of this enzyme.[24] Criddle has refined this observation by finding that chloramphenicol preferentially inhibits the synthesis of the large subunit while cycloheximide preferentially inhibits formation of the small subunit.[25] This implies that the large subunit is synthesized on a chloroplast ribosome and the small subunit on a cytoplasmic ribosome.

The large size of ribulose diphosphate carboxylase gives it a distinctive appearance in electron micrographs as a 100 angstrom particle of regular dimensions.[26] This property allows the tentative identification of ribulose diphosphate carboxylase on one of the chloroplast membrane surfaces as viewed in the electron microscope.[27]

Wishnick and Lane found that the enzyme activity is cyanide sensitive.[28] Next, a tightly bound copper was found in the ribulose diphosphate carboxylase which is presumably the site of cyanide inhibition and part of the active site.[29]

There is kinetic evidence that magnesium ion may act as an allosteric activator of spinach ribulose diphosphate carboxylase.[30] Bassham has found that the pH optimum of crude enzyme preparations is shifted from 8.5 to 7.6 by Mg^{+2}.[31]

A divergence in size or ribulose diphosphate carboxylases has been revealed by studies of the enzyme from photosynthetic bacteria and algae

TABLE 4.1   THE SIZE OF RIBULOSE DIPHOSPHATE
CARBOXYLASE FROM VARIOUS PLANTS

		Approximate Molecular Weight
Purple non-sulfur bacteria	*Rhodospirillum rubrum*	83,000
	Rhodopseudomonas spheroides	240,000
Purple sulfur bacteria	*Chromatium* D	600,000
Blue-green algae	*Anacystis nidulans*	600,000
Green algae	*Chlorella pyrenoidosa*	600,000
	Euglena gracilis	600,000
Higher plants	Spinach	600,000
	Spinach beet	600,000
	Chinese cabbage	600,000

(Table 4.1).[32,33] The size and general characteristics of ribulose diphosphate carboxylase in *R. rubrum* and *Rps. spheroides* are not altered by growing these organisms in heterotrophic, semiautotrophic, or autotrophic conditions and all of the bacterial enzymes require magnesium ion for activity.[34] Akazawa et al. found that the *Chromatium* enzyme is inhibited by an antibody prepared against spinach ribulose diphosphate carboxylase,[33] further emphasizing the evolutionary jump among the bacteria.

Finally, it may be noted that ribulose diphosphate carboxylase has been isolated from a heterotrophic culture of the non-photosynthetic facultative autotroph *Hydrogenomonas*.[35] This enzyme has a molecular weight of about 500,000 and appears to consist of identical subunits. All of these data suggest that a path of evolution will be traced out among the microorganisms when sufficient molecular data are available.

4.4   PHOTORESPIRATION

In addition to catalyzing the reaction for which it is named, ribulose diphosphate carboxylase is able to catalyze a cleavage of ribulose diphosphate with oxygen to yield phosphoglycerate and phosphoglycolate.[36] Oxygen inhibits photosynthesis and stimulates respiration in many plants. Ogren and Bowes found that oxygen was a competitive inhibitor of carbon dioxide incorporation catalyzed by ribulose diphosphate carboxylase and the kinetics of this inhibition were similar to the kinetics of inhibition by oxygen of photosynthesis in intact leaves.[37,38] Lorimer et al. have confirmed these observations and extended them by showing a stoichiometry of two atoms of oxygen consumed per molecule of ribulose diphosphate cleaved leading to the proposal that this activity be called the ribulose diphosphate oxygenase reaction.[39]

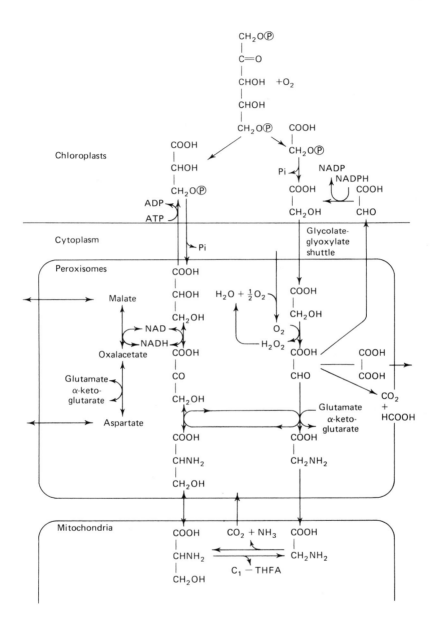

Figure 4.3   A postulated reaction sequence relating photorespiration to the metabolic activities of peroxysomes and mitochondria. [Modified from Tolbert, N. E., *Ann. Rev. Plant Physiol.,* **22**, 45 (1971).]

This discovery clarified a long-standing puzzle in the understanding of photosynthetic carbon metabolism. The phenomenon of photorespiration has been recognized for some time.[40] In many plants, light stimulates oxygen consumption and carbon dioxide production. Photorespiration is generally obscured by the oxygen production and carbon dioxide fixation of photosynthesis but may be of physiological importance when it prevents net carbon fixation at low concentrations of external carbon dioxide. Zelitch has championed the role of glycolate as a substrate for photorespiration since either genetic or environmental manipulations that increase photorespiration increase glycolate metabolism.[41] Feeding glycolate to leaves in the dark will cause respiration. Conversely, selective inhibition of glycolate oxidation by alpha hydroxypyridine methane sulfonate increases net carbon dioxide fixation and eliminates photorespiration in leaves.[42] Isotope labelling data indicated that glyoxylate came from carbons one and two of ribulose diphosphate, and now the oxygen dependent cleavage of that substrate by ribulose diphosphate carboxylase makes the origin of glycolate clear. There is a specific phosphoglycolate phosphatase in the chloroplast that converts the phosphoglycolate to glycolate and may be involved in transport of glycolate to the outside of the chloroplast as suggested in Figure 4.3.[43]

Subsequent metabolism of glycolate must occur in a microbody called the peroxysome.[44] Here glycolate is oxidized to glyoxylate by the flavoprotein glycolate oxidase with concomitant production of hydrogen peroxide. This enzyme might further oxidize the glyoxylate to oxalate. Catalase that is present within the peroxysome should dispose of the hydrogen peroxide. Glyoxylate can be converted to carbon dioxide and formate by hydrogen peroxide and a chloroplast system catalyzing this reaction has been described.[45] A transaminase is available to convert glyoxylate to glycine using glutamic acid as the amino group donor. Alternately, the glyoxylate might be returned to the chloroplast where it can be reduced to glycolate at the expense of reduced pyridine nucleotide. A complex series of reactions can interrelate glycolate to glycerate involving both peroxysomes and mitochondria. The oxidation of glycolate to glyoxylate allows for the disposal of reducing power generated in the light. Low carbon dioxide *in situ* would lead to the splitting of ribulose diphosphate by oxygen. The accumulation of reduced pyridine nucleotides could then be relieved by glycolate oxidation to prevent photodestruction of the chloroplast which would occur if the absorbed radiant energy could not be discharged in controlled chemical work.

Peroxysomes are microbodies found in all higher plants and their appearance in the cell parallels the development of photosynthetic competence.[46,47] The peroxysomes appear to be very close to the chloroplasts when seen in electron micrographs of cell sections. Peroxysomes are not found in algae and these organisms appear to excrete glycolate. Although *Chlorella* contain enzymes for the conversion of glycolate to glycerate,[48] it is by no means clear that glycolate is related to photorespiration in algae.[49]

4.5   THE C_4 PATHWAY OF CO_2 FIXATION

A group of plants called tropical grasses which includes corn, sorghum, sugar cane, etc., possess a unique elaboration of the Calvin cycle. These plants show a much higher affinity for carbon dioxide and do not lose carbon dioxide in the light by photorespiration. When these plants are exposed to radioactive carbon dioxide in the light, then extracted and the early fixation products chromatographed, the first product of fixation appears to be a C_4 dicarboxylic acid — malate, aspartate, or oxaloacetate — then the label moves to the C_1 position of phosphoglyceric acid and subsequently follows the conventional Calvin cycle.[50] These plants have a very active phosphoenolpyruvate carboxylase as well as an interesting enzyme called pyruvate phosphate dikinase.[51,52,53,54]

$$\text{Pyruvate} + \text{PO}_3^{=} + \text{ATP} \xrightarrow[\substack{\text{pyruvate} \\ \text{phosphate} \\ \text{dikinase}}]{\text{Mg}^{+2}} \text{PEP} + \text{AMP} + \text{PP}_i$$

This enzyme catalyzes the incorporation of inorganic phosphate into the pyrophosphate that is produced in the reaction, hence the name dikinase. There are interesting experiments on the photocontrol of this system in corn — the pyruvate phosphate dikinase is activated by light and disappears in darkness.[55] Evidence has appeared to suggest that this photocontrol may be exerted through the reversible oxidation and reduction of adjacent thiols on the enzyme. The adenylate kinase and pyrophosphatase that serve in the cycle are phytochrome induced enzymes.[56]

Phosphoenolypyruvate carboxylase is available to fix carbon dioxide into oxaloacetate. The enzyme is inhibited by its product oxaloacetate at or near physiological concentrations that would prevent the accumulation of oxaloacetate when there is insufficient NADPH to drain it off.[57] Neither malate nor aspartate inhibits the phosphoenolpyruvate carboxylase of tropical grass leaves although these compounds are feedback inhibitors in other types of cells. The phosphoenolypyruvate carboxylase activity in corn is not affected by oxygen concentration while the ribulose diphosphate carboxylase is prevented from catalyzing carbon dioxide fixation by oxygen.[38] Thus the tropical grasses may be insensitive to oxygen induced photorespiration since their main entry point for fixing carbon is phosphoenolpyruvate carboxylase. In this connection it is interesting to note that plants using ribulose diphosphate carboxylase as the primary site of carbon dioxide fixation show a more pronounced discrimination against the heavy isotope of carbon than do tropical grasses.[58]

Once fixed in oxaloacetate, the carbon may accumulate in an intermediary pool of malate or aspartate as shown in Figure 4.4.[59] For subsequent metabolism the oxaloacetate presumably decarboxylates via phosphoenol-

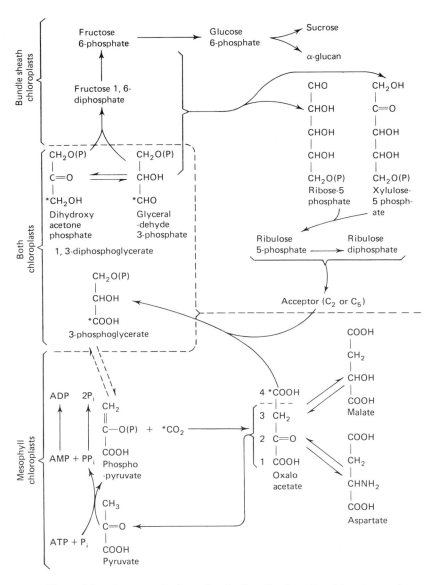

Figure 4.4 A proposed scheme for the C_4 – dicarboxylic acid pathway of photosynthesis. [Redrawn from Hatch, M. D. and C. R. Slack, *Ann. Rev. Plant Physiol.*, **21**, 141 (1970).]

pyruvate carboxylase or pyruvate carboxykinase to give CO_2 which is used by the ribulose diphosphate carboxylase.[60] Since the ribulose diphosphate carboxylase of corn will catalyze the oxygenase reaction, one must look beyond the isolated enzymes for an explanation of absence of photorespiration in these plants. Leaves of tropical grasses show two distinct types of chloroplast-containing cells. Mesophyll cells contain chloroplasts with the usual grana stacks and the C$_4$ cycle enzymes enumerated above, but they ·do not contain starch. Bundle sheath cells have chloroplasts that do not show grana stacks, but they do contain starch and have a conventional complement of Calvin cycle enzymes. It seems likely that malate or aspartate is formed in the mesophyll and then transported to the bundle sheath cells where these acids serve as a source of carbon dioxide via decarboxylations (Fig. 4.4). Within the bundle sheath cell, the carbon dioxide is evidently protected from loss via photorespiration despite the abundance of peroxysomes in these cells. This system is expensive since direct fixation by the Calvin cycle demands three ATP and two NADPH per carbon dioxide fixed while the C$_4$ cycle requires an additional two ATP. This expenditure allows about a one hundredfold increase in affinity for carbon dioxide by the plant, but the mechanism by which this increase in affinity for carbon dioxide is achieved is unknown.

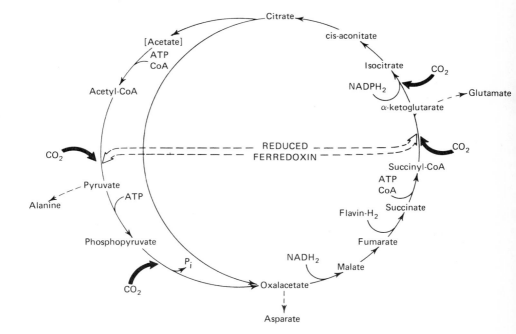

Figure 4.5   The reductive carboxylation cycle. One turn of the complete cycle (represented by the one-sided ellipse) results in the incorporation of four molecules of CO$_2$. One turn of the short cycle (represented by the circle) results in the incorporation of two molecules of CO$_2$. [From Evans, M. C. W., B. B. Buchanan, and D. I. Arnon, *Proc. Natl. Acad. Sci. USA,* 55, 928 (1966).]

4.6 THE REDUCTIVE CARBOXYLATION CYCLE

Arnon has elucidated another path of CO_2 fixation that operates in some photosynthetic bacteria.[61] The sequence involves the net fixation of four moles of CO_2 to produce a mole of oxaloacetate and requires two new carboxylation reactions of acetyl and succinyl CoA to pyruvate and alpha ketoglutarate respectively (Fig. 4.5). These reductive carboxylations require ferredoxin. These reactions have been demonstrated in cell-free preparations of *Chlorobium thiosulfatophilum* and *Chloropseudomonas ethylicum*.[62,63] *Rhodopseudomonas palustris* may also have this pathway.[64] In addition, *Chromatium* contains an enzyme that catalyzes the reductive carboxylation of phenylacetyl CoA to phenylpyruvate using reduced ferredoxin and thiamine pyrophosphate as coenzymes.[65] Thus reductive carboxylations with reduced ferredoxin as the electron donor may be of general biosynthetic utility in certain photosynthetic bacteria.

GENERAL REFERENCES

Hatch, M. D., and C. R. Slack. "The C_4 Dicarboxylic Acid Pathway of Photosynthesis," *Progress in Phytochemistry*, 2, 35 (1970).

Hatch, M. D., C. B. Osmond, and R. O. Slayter (eds.). *Photosynthesis and Photorespiration*. New York: Wiley-Interscience, 1971.

Tolbert, N. E. "Microbodies — Peroxisomes and Glyoxysomes," *Ann. Rev. Plant Physiol.*, 22, 45 (1971).

Zelitch, I. *Photosynthesis, Photorespiration and Plant Productivity*. New York: Academic Press, 1971.

5 ‖ HEXOSE ASSIMILATION AND THE CELL WALL

Hexoses may be generated via photosynthetic CO_2 fixation or by the conversion of lipid through the glyoxylate cycle. Aside from their consumption via glycolysis or the shunt pathway, the hexoses have an interesting metabolism in the generation of both metabolic reserves and structural elements.[1],[2]

5.1 REACTIONS OF HEXOSES

Below is a brief survey of the types of reactions involved.

Phosphorylases.

These enzymes catalyze the following type reaction:

$$\text{Hexose} - \text{P} + \text{acceptor} \rightleftharpoons \text{hexose} - \text{acceptor} + \text{P}_i$$

57

There are phosphorylases for sucrose (bacterial), starch, glycogen, and laminarabinose. It appears that the phosphorylases work in the cell to produce hexose phosphate by degradation of polysaccharides.

Inulin Synthesis.

In artichoke, inulin synthesis is an exceptional process in that this fructose polymer is built with monomers from the end of a trisaccharide which come in turn from a disaccharide.

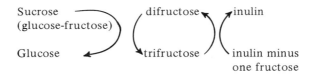

Nucleoside Diphosphate Sugars.

These compounds constitute the main activated form for much of hexose utilization.

$$\text{Nucleoside triphosphate + sugar 1 phosphate} \xrightleftharpoons{\text{pyrophosphorylase}}$$
$$\text{nucleoside diphosphate - sugar + PP}_i$$

In plant extracts, one finds many kinds of sugars stuck on various nucleoside diphosphates and the respective pyrophosphorylases. UDP sugars appear to be most popular in metabolic circles, but ADP-glucose in starch synthesis and GDP-glucose in cellulose synthesis are interesting special cases.

Epimerization.

Epimerase reactions are a convenient way to interconvert sugars attached to the nucleoside diphosphates.

$$\text{UDP-glucose} \rightleftharpoons \text{UDP-galactose}$$

$$\text{UDP-xylose} \rightleftharpoons \text{UDP-arabinose}$$

$$\text{UDP-glucuronic acid} \rightleftharpoons \text{UDP-galacturonic acid}$$

These epimerase reactions can be demonstrated in plant extracts. The epimerase mechanism is still obscure. Although the enzyme from several sources contains bound NAD, it is not clear how this coenzyme participates in the reaction.

Oxidation.

Nucleoside diphosphate sugar is oxidized at the hexose carbon number six.

$$UDP\text{-glucose} \xrightarrow[2\ NADH]{2\ NAD} UDP\text{-glucuronic acid}$$

The obvious aldehyde intermediate has yet to be trapped. There are scattered reports of oxidation at C_1 and C_6 of free sugars in plant extracts — usually by a flavoprotein similar to glucose oxidase.

Cyclization.

Glucose 6 phosphate is converted to glucuronic acid by cyclization to myoinositol followed by ring opening.[3] The first enzyme of this sequence has been purified[4] and subsequent steps (Fig. 5.1) are known from isotope studies.[5]

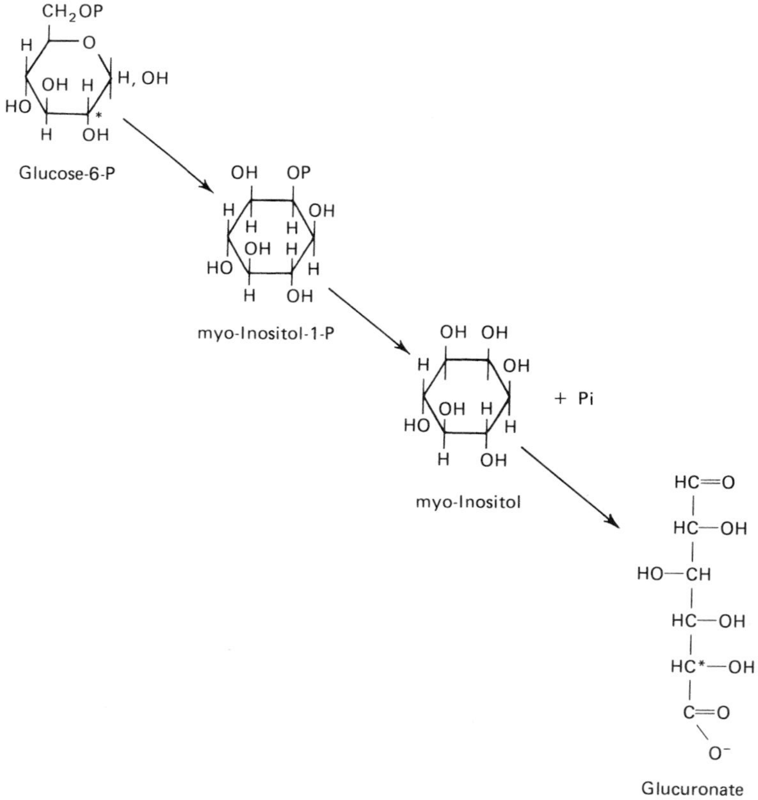

Figure 5.1 The conversion of glucose-6-phosphate to glucuronate via cyclization.

Decarboxylation.

UDP-glucuronic acid but not UDP-galacturonic acid is decarboxylated in bean and wheat-germ preparations.

$$\text{UDP-glucuronic acid} \longrightarrow CO_2 + \text{UDP-xylose}$$

Parsley, grown as a cell suspension culture, yields two distinct enzymes that catalyze the formation of UDP-xylose from UDP-glucuronic acid.[6] One of these enzymes requires NAD for its catalytic activity and its appearance in the culture is light induced. As with other isozymes, one presumes that each form of the catalyst provides UDP-xylose for a different use. Different uses may require separate regulatory mechanisms which in turn require separate isozymes. Since the UDP-xylose can then be epimerized to UDP-arabinose, this decarboxylation provides two pentoses as nucleoside diphosphate derivatives properly prepared for polymerization into pentosans.

Oxidation, Reduction, and Epimerization.

These reactions account for the appearance of rhamnose in plant material. These reactions are described by Figure 5.2.

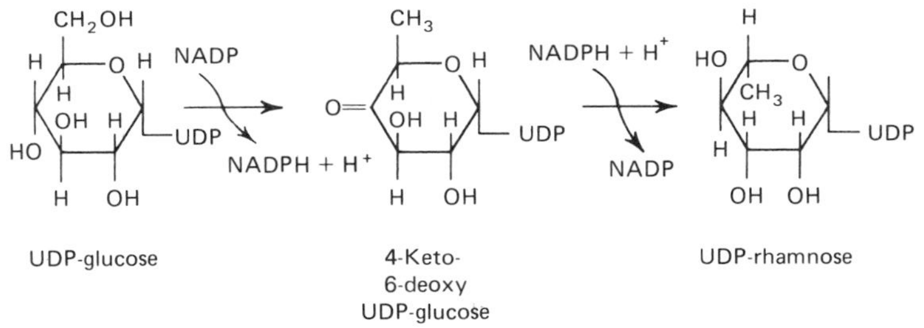

| UDP-glucose | 4-Keto-6-deoxy UDP-glucose | UDP-rhamnose |

Figure 5.2   The oxidation and reduction sequence which epimerizes UDP-glucose to UDP-rhamnose.

Elimination and Rearrangement.

The production of apiose, a branched chain pentose found in parsley, involves elimination and rearrangement. The mechanism of production from glucose is still at the radioactive carbon stage[7] but the reaction evidently involves the loss of C_6 and the branching out of C_3 to give the hydroxymethyl group[8,9,10] as outlined in Figure 5.3. An enzyme has been purified from *Lemna minor* which catalyzes the synthesis of both UDP-apiose and UDP-xylose from UDP-glucuronic acid.[11]

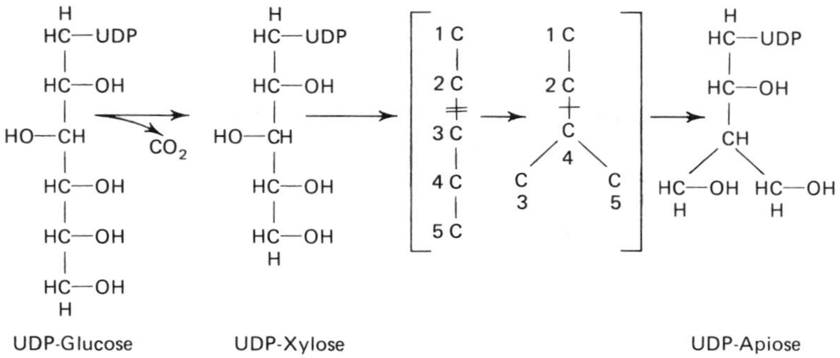

Figure 5.3   The conversion of UDP-glucose to UDP-apiose.

Glycoside Formation.

This is a profligate activity of plant tissues in that sugars are tacked on to all manner of phenolic compounds, flavonoids, alkaloids, and sterols. The glucosylation reactions that follow are known in plant materials and demonstrate the general utility of UDPG as a hexose donor.

1. UDP-glucuronic acid + o-aminophenol $\longrightarrow$
   o-aminophenol-beta, D-glucuronic acid + UDP
2. UDP-glucose + hydroquinone $\longrightarrow$ hydroquinone glucoside + UDP
3. UDP-glucose + phenyl-beta glucoside $\rightleftharpoons$
   UDP + phenyl-beta gentiobiose
4. UDP- (or TDP-) glucose + quercetin $\longrightarrow$ 3 quercetin beta D
   glucoside + UDP (or TDP)
   $\downarrow$ UDP (or TDP) rhamnose
   rutin

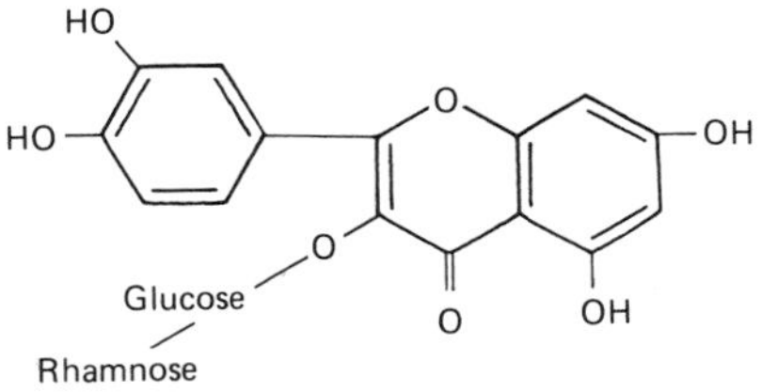

Figure 5.4   Rutin, a flavone glycoside.

5.2   SUCROSE SYNTHESIS

Disaccharide synthesis in higher plants must begin with sucrose since this is the most frequently encountered and quantitatively most abundant disaccharide.

Leloir's work has indicated two separate routes for sucrose synthesis.

1. UDP-glucose + fructose $\rightleftharpoons$ sucrose + UDP
2. UDP-glucose + fructose 6 phosphate $\rightleftharpoons$ sucrose phosphate + UDP

One finds very little sucrose phosphate in plants, but (or perhaps because) one finds plenty of sucrose phosphate phosphatase. The irreversible hydrolysis of the phosphate ester could be a pulling force in tissue like beet in which much sucrose accumulates despite the reversibility of the two reactions from UDP-glucose. A second evidence that reaction (2) is the physiological path of sucrose synthesis comes from feeding experiments. C^{14} glucose appears in the fructofuranosyl part of sucrose much more rapidly and in greater concentration than in free fructose (Fig. 5.5) indicating that free fructose is not an intermediate in the synthesis of sucrose. There is plenty of sucrose phosphate synthetase present[12] – although it must be measured in the presence of EDTA to inhibit the sucrose phosphate phosphatase – and the synthetase activity level is high enough to account for the rate of sucrose synthesis by the tissue. A sucrose phosphate synthetase has been purified from bean cotyledons and it shows a complex pattern of activation and inhibition.[13]

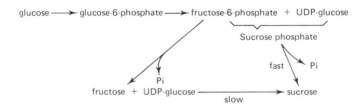

Figure 5.5   Sucrose synthesis.

Why then have the reaction (1) that forms sucrose from fructose and UDPG? Quite plausibly this enzyme is present in order to move glucosyl units from sucrose into starch. This happens in many storage organs and Cardini has suggestive evidence for this role in that the sucrose synthetase (but not the sucrose phosphate synthetase) is inhibited by phenolic glycosides (feedback regulators?) and uses ADP-glucose as well as UDP-glucose (starch formation proceeds through either but especially ADP-glucose)[14] (Fig. 5.6).

The discovery of several types of sucrose synthetase suggests other roles for this enzyme. A sucrose synthetase has been found in bean seedlings and this enzyme uses many different nucleoside diphosphate derivatives of glucose as precursors of sucrose.[15] Two isozymes of sucrose synthetase have been reported with the suggestion that one might catalyze synthesis and the other cleavage of sucrose.[16]

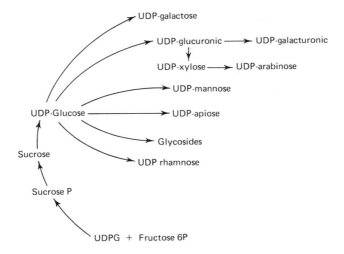

Figure 5.6   A summary diagram of reactions of UDP sugars.

5.3   GALACTINOL AS A GALACTOSE DONOR

Raffinose is sucrose with galactose attached as an alpha glucoside to C_6 of glucose and probably is formed via

$$\text{UDP-galactose + sucrose} \rightleftharpoons \text{raffinose + UDP}$$

Stachyose is a tetrasaccharide that is a galactosyl derivative of raffinose and here alpha D galactopyranosyl (1-1) myoinositol (called galactinol) is the galactose donor to raffinose. The sequence in *Phaseolus* is[17]

$$\text{UDP-galactose + myoinositol} \rightleftharpoons \text{galactinol + UDP}$$

$$\text{galactinol + raffinose} \rightleftharpoons \text{stachyose + myoinositol}$$

which in *Vicia* can recycle one further:

$$\text{Stachyose + galactinol} \rightleftharpoons \text{verbascose + myoinositol}$$

There is one more cycle to ajugose.

$$\text{Verbascose + galactinol} \rightleftharpoons \text{ajugose + myoinositol}$$

These reactions, with the synthesis of inulin, provide two exceptions to the nearly complete hegemony of the nucleoside diphosphate sugars in linking sugars together.

5.4   STARCH SYNTHESIS

Starch is the major carbohydrate reserve in higher plants. Chemically, it is a homogeneous polymer containing only glucose units which in amylose are alpha 1,4 linked while in amylopectin occasional alpha 1,6 links give a branched structure. Certain waxy varieties of grain have only amylopectin.

X-ray diffraction shows the chain to be helical with six glucose residues per turn. When starch is subjected to the starch iodine test, iodine molecules sit in the center of each gyre of the helix and in the long helices, the iodine-iodine interactions cause the maximum wavelength in the iodine spectrum to shift to longer wavelength.

Phosphorylase falls in and out of popularity as a catalyst for starch synthesis, but like muscle phosphorylase, its main role is assumed to be in starch breakdown.

$$\text{Glucose 1 phosphate + primer} \rightleftharpoons \text{Pi + starch}$$

This enzyme has been crystallized from potato. It is alpha 1,4 linkage specific, molecular weight 207,000, and contains two moles of pyridoxal phosphate. It has no serine phosphate analogous to the residue in muscle phosphorylase which is the site of action for muscle phosphorylase kinase and phosphorylase phosphatase. Needless to say, there is no cyclic AMP activation of the potato enzyme.

The tendency to assign phosphorylase a role in plant starch synthesis arises from the fact that (1) phosphorylase activity rises in tissues that are actively synthesizing starch, as in ripening grain and (2) some phosphorylases will polymerize glucose with little or no primer[18] in contrast to most starch synthesizing mechanisms that require primer. Thus some phosphorylases might act to provide primers for the other systems. Tsai and Nelson found four phosphorylases in ripening corn.[19] Two of these enzymes require no detectable primer. These enzyme levels are reduced in the "shrunken 4" mutation in which starch synthesis is seriously diminished.

A strong argument against phosphorylase as a major route of starch synthesis comes from the high concentration of inorganic phosphate in the plant cell creating an unfavorable equilibrium for starch synthesis, but compartmentation might circumvent this difficulty.[20]

Leloir et al. found a system for incorporating glucose into starch from UDP-glucose. This is an insoluble enzyme system found in many plants and associated with the starch grains. Next, an enzyme that incorporates glucose into starch from ADP-glucose turned up and can be prepared in soluble form from sweet corn. The ADP-glucose-starch transglucosylase is widely distributed[21] and the soluble enzyme may be distinct from an enzyme catalyzing the same reaction but bound to starch grains. Distinct isozymes of ADP-glucose-starch transglucosylase have been found in several tissues. In corn, an isozyme is present that polymerizes glucose without primer.[22] In rice, the isozymes differ in their preference for the length of the glucosyl acceptor.[23]

The UDP-glucose glucosyl transferase is much more abundant than the ADPG enzyme, but the UDP enzyme has a much lower turnover. Which is the main route to starch — via ADPG or UDPG? Nelson has reported on a corn mutant that lacks ADP-glucose pyrophosphorylase,

$$\text{glucose 1 phosphate} + \text{ATP} \rightleftharpoons \text{ADP-glucose} + \text{pyrophosphate}$$

and this mutant shows only 25% of the wild type starch level.[24,25] Akazawa found that iodoacetate would inhibit starch deposition in ripening rice grains. Adenine nucleotides and ADP-glucose protect the starch synthesis by the rice grains from inhibition by iodoacetate which is another indication that the ADP-glucose path is the main physiological route.[26] Perez et al. have described a synthetase bound to starch granules in rice and this enzyme activity parallels the

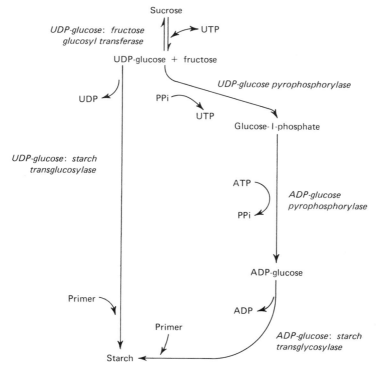

Figure 5.7   Conversion of sucrose to starch via nucleoside diphosphate sugars.

cell activity in starch synthesis.[27] This enzyme used ADP-glucose but not UDP-glucose as a glucosyl donor.

The UDP-glucose enzymes might be involved in starch formation from sucrose. Figure 5.7 outlines the possible interrelations of sucrose and starch. The synthesis of starch by corn mutants with low levels of ADP-glucose pyrophosphorylase is most likely accomplished through transfer of glucosyl units from sucrose via UDP. In rice, the incorporation of C^{14} glucose from sucrose to starch is stimulated by ATP while sucrose breakdown is promoted by UDP. Regulation of the sequence of reactions leading to starch should reasonably occur at ADP-glucose pyrophosphorylase and this had been found.

$$\text{Glucose 1 phosphate } + \text{ ATP} \xrightleftharpoons{\text{pyrophosphorylase}} \text{ADP-glucose } + \text{PP}_i$$

$$\text{ADP-glucose } + \text{ primer} \xrightleftharpoons{\text{transglucosylase}} \text{ADP } + \text{ glucosyl-primer}$$

Preiss has purified the pyrophosphorylase from a number of plant tissues and finds the following: phosphoglyceraldehyde and phosphoglyceric acid stimulate 5 to 15 fold; fructose 6 phosphate and fructose 1,6 diphosphate stimulate 2 to 6 fold; Pi and ADP inhibit and are reversed by phosphoglyceraldehyde.[28,29,30] Therefore, a rise in the phosphoglyceric acid, phosphoglyceraldehyde, or hexose phosphate level caused by photosynthesis brings an increase in starch synthesis by stimulating the ADP-glucose pyrophosphorylase activity. High levels of inorganic phosphate and ADP, caused by lack of photophosphorylation or an energy deficit in the cell, might reduce incorporation of glucose into starch. However, changes in the ADP and inorganic phosphate pool sizes are yet to be documented.

The conversion of amylose to amylopectin is accomplished by the Q enzyme which is an alpha 1,4 $\longrightarrow$ alpha 1,6 amylo transglucosidase. The branching enzyme for synthesis of phytoglycogen has recently been purified from maize endosperm.[31] The Q enzyme of potato uses only large segments – 50 glucose residues – as either donor or acceptor.[32]

In general, starch is first synthesized in the chloroplast from newly fixed carbon. Then sugar may be exported as sucrose to either the growing region or the storage organ where starch is synthesized in a leucoplast.

The starch grains have identifiable microscopic structure that is characteristic of the species of origin – a layering visible as rings of varying density under the light microscope. This layering of starch in barley and wheat endosperm results from the variation in starch deposition in day vs. night since plants grown in continuous light show no layering. In contrast, potato and tobacco starch is layered even in continuous light suggesting that starch formation is controlled by an endogenous circadian rhythm.

5.5   STARCH BREAKDOWN

This process is easily observed in germinating seeds in which there is massive conversion of reserve starch to carbohydrate. The starch degrading enzymes appear in the germinating seed either through hormone induced *de novo* synthesis[33] or by liberation of a proenzyme from a zymogen granule followed by proteolysis to give an active enzyme.[34]

In rice, the endosperm starch goes down on the fourth day after germination which is about the time that the amylase activity appears. Phosphorylase activity at this point in time would give only 5% of the starch hydrolysis compared to the amylase activity.[35] In germinating peas, alpha amylase is a major factor in the 6- to 12-day period after germination; then phosphorylase and beta amylase participate in further hydrolysis.[36] One level of regulation of starch hydrolysis may be achieved by differential binding of enzymes to the starch granule. This is suggested by studies of phosphorylase absorption to amyloplasts.[37] Absorption of the enzyme to the starch granule substrate is blocked by high hexose levels and when the hexose level is decreased, the enzyme is then absorbed to the starch granule where it can liberate more hexose.

Beta amylase has been studied in many plants and crystallized from sweet potato. It attacks starch from the reducing end, hydrolyzing alternate alpha $1 \longrightarrow 4$ links to give the disaccharide maltose. Catalysis is stopped by the 1,6 branch points in amylopectin so that beta amylase can digest only the outer chains of this molecule.

Alpha amylase is abundant in germinating cereal, has a molecular weight of 60,000 and requires calcium ion for activation and stabilization. This enzyme first splits out oligosaccharides (dextrins) from both amylose and amylopectin and then splits these into maltose with a little glucose and maltotriose from the branch points.

Isoamylase in rice and the R enzyme in corn, bean, and potato split alpha $1 \longrightarrow 6$ linkages found at the branch points.[38]

Alpha glucosidases are present in many tissues to convert disaccharides to glucose.

The sequential appearance and the localization of enzymes in the pea seedling indicate the sequence of enzyme activities shown in Figure 5.8.[39] An alternate sequence in the early cotyledon might be via phosphorylase that appears earlier and the hydrolytic enzymes that appear later. In the pea seed, an amylopectin-1,6 glucosidase appears on germination. The proenzyme is first released from a zymogen granule and then activated by proteolysis.[40] In potato starch, degradation may be regulated by changes in the permeability of amyloplast membrane that surrounds the starch granules.[41]

Occasionally starch is replaced by another polysaccharide as the carbohydrate reserve in plant tissue. Inulin, the polyfructose formed by an unusual

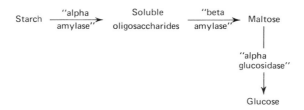

Figure 5.8   The degradation of starch to glucose.

transfructosylation from trifructoside, is the reserve in artichoke. A polyfructose is formed in dahlia presumably through UDP-fructose. Mannans are present in a few nut tissues and in yeast in which they arise via GDP-mannose. In the algae, variations on polyglucose, for example, laminarin and Floridean starch (alpha 1 ⟶ 3 with beta 1 ⟶ 6 and alpha 1 ⟶ 4 branch points), are known.

In the blue-green algae, there are several forms of polysaccharide. An interesting report of incorporation of acetate into poly-beta-hydroxybutyrate by a blue-green alga[42] heightens the similarity of these algae to photosynthetic bacteria that use poly-beta-hydroxybutyrate as their main carbon and energy reserve.

5.6   CELL WALL POLYSACCHARIDES

Cellulose.

Cellulose is the ultimate product of about one-third of all the carbon dioxide fixed by the plant. It exists as a beta 1 ⟶ 4 linked glucose polymer without branching and of enormous length and negligible solubility. Although generally considered a homogeneous polymer, there is evidence for the occasional appearance of units other than glucose. The cellulose chains are organized in crystalline microfibrils that may account for up to 20% of the cell wall volume. One can extract all of the other wall constituents and cause remarkably little change in the form or mechanical properties of the cell. These microfibrils prevent an increase in girth of the cell but do allow cell elongation.[43] The microfibrils are laid down on the inner surface of the cell's primary wall parallel and perpendicular to the major axis of the cell. The coil acts like a barrel hoop restraining the cell like a coil spring that can be extended in length as the cell elongates. In electron micrographs, one can follow the microfibril rearrangement with the coil spreading and the pitch going from perpendicular to parallel to the growth axis.

Synthesis of cellulose by a cell-free enzyme preparation was first achieved with the bacterium *Acetobacter xylinium* using UDP-glucose as precursor. In

1964, Elbein reported on an enzyme preparation from a higher plant (*Phaseolus*) that would incorporate glucose from GDP-glucose into a cellulose-like product. The activity was particulate, unstable, and difficult to purify. The glucosyl transferase could not be separated from the glycosyl acceptor, i.e., cellulose in the preparation. The activity was stimulated by cobalt, manganese, and magnesium ions. A similar preparation from lupine uses UDP-glucose for cellulose synthesis. The evidence for the role of lipid in higher plant cellulose synthesis is controversial. An enzyme synthesis of cellulose-like polysaccharides by an oat extract uses UDP-glucose to form glucolipid and polysaccharide.[44] It is not clear that the glucolipid is an intermediate in the polysaccharide synthesis. Phospholipase gives occasional inhibition of polysaccharide synthesis. Detergents inhibit and lecithin reverses detergent inhibition, but this latter evidence might apply equally well in non-lipid requiring systems. The preparations using UDP-glucose form beta 1 $\longrightarrow$ 3 as well as beta 1 $\longrightarrow$ 4 linkages which is disconcerting since beta 1 $\longrightarrow$ 3 linkages are not found in cellulose. Hassid's laboratory has recently reported on a soluble enzyme system from either bean or lupine that uses GDP-glucose for synthesis of cellulose.[45,46] The enzymes are solubilized by digitonin and show no evidence for glycolipid involvement.

Degradation of cellulose is rarely done by plants. A few germinating seeds will degrade cellulose with cellulase to the disaccharide cellubiose which is split by cellubiase to glucose. Many bacteria hydrolyze cellulose by the same route (intestinal flora of termites are uniquely talented in this regard).* The hepato-pancrease juice of the snail is a currently popular source of cellulase for preparing protoplasts from yeast.

Non-Cellulosic Polysaccharides.

Cellulose is embedded in an amorphous matrix which in the young or primary wall is mainly polysaccharide, but in the secondary wall is polysaccharide plus a great deal of lignin. The view that homopolysaccharides — xylans, arabans, glucurans, and galacturans — exist in the cell wall as independent units is probably incorrect in that extraction of these polymers seems to involve hydrolysis of covalent bonds which at least occasionally link them together.

Villemez has studied the synthesis of several of these non-cellulose poly-saccharides in higher plant preparations. He has developed evidence for a lipid soluble precursor of cell wall polysaccharide using both labelled UDP-glucuronate and GDP-mannose as precursors.[47] An isoprene carrier analogous to the one used in some bacterial cell wall polysaccharide synthesis is indicated as

---

*"A primal termite bit on wood,
 Tasted it and found it good,
 And that is why your cousin May
 Fell through the parlor floor today."

Ogden Nash, "The Termite" (from *Verses from 1929 On*)

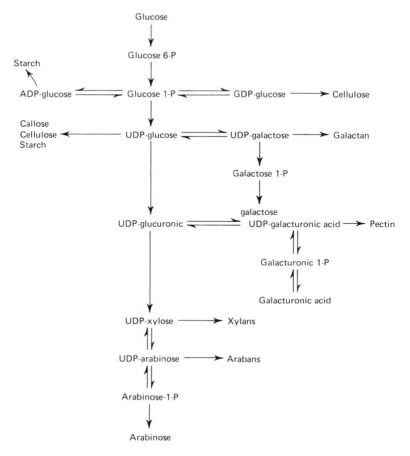

Figure 5.9   Glucose conversion into various polysaccharides via nucleoside diphosphate sugars.

the functional lipid in these plant preparations and there is evidence for an oligosaccharide-protein precursor to the final polysaccharide product.

Polygalacturonic acids are alpha 1 ⟶ 4 linked backbones with rhamnose, galactose, arabinose, and 2-0-methyl xylose as possible side chains. Many of the carboxylic groups are esterified to methanol and Hassid has found a cell-free preparation from bean tissue that catalyzes this esterification using S-adenosyl methionine as methyl donor.[48,49] Methylation occurs after polymerization. One assumes that the main chain is built with UDP-galacturonic acid residues, and a cell-free preparation catalyzing such a reaction is being characterized. A variety of "pectic" degrading enzymes are known in plant pathogens.

There are polyxylans, arabans, etc. that are probably all heteropolymers with other sugars as side chains. Cell-free synthesis is known only in fragments[50] and might best be summarized by Figure 5.9.

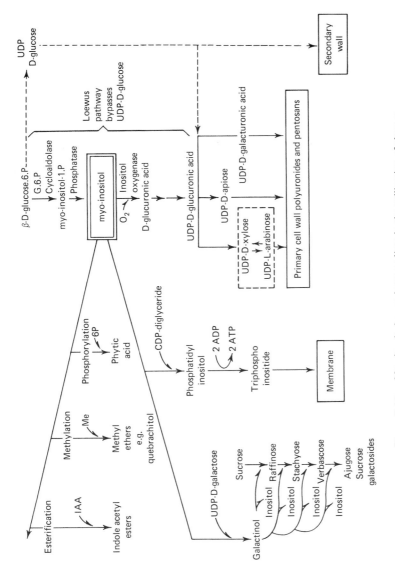

Figure 5.10 Myo-inositol as an intermediate in the utilization of glucose. [From Lamport, D. T. A., *Ann. Rev. Plant Physiol.*, **21**, 235 (1970).]

In addition to the UDPG pathways, a path to polysaccharides through myoinositol (summarized in Fig. 5.10) seems operative in many plant tissues.[51,52] This may be a route to make polyuronides and pentosans that bypasses a control of UDP-glucose. Thus the cell might continue to synthesize pectins, xylans, and arabans after glucan synthesis has been shut down. Northcote has published a study of general interest on the subcellular origins of cell wall polysaccharides.[53] Tritiated glucose was fed to root tips which were then autoradiographed. There was an apparent accumulation of a pectin polysaccharide in the Golgi apparatus which then moved through the cytoplasm and across the plasmalemma to the cell wall. The same conclusion was reached by *in vitro* studies.[54] A fraction rich in Golgi-like structures was isolated from pea seedling roots and was shown to be the site of synthesis of pectic substances and hemicellulose.

Since knowledge of the synthesis of cell wall polysaccharides is sketchy, the regulation of these processes is unknown. Van der Woude et al. have observed that the auxin 2,4 dichlorophenoxyacetic acid stimulates polysaccharide synthesis both *in vivo* and in plasma membrane fragments isolated from onion.[55] Hormonal effects on the plant cell wall have been obvious for a very long time. One hopes that these effects can be interpreted in molecular terms as the polysaccharide metabolism in the wall is worked out.

5.7   LIGNIN

By way of review, recall that the primary wall of a plant cell consists of cellulose — a glucose homopolymer of enormous chain length. The chains are packed in microfibrils which in turn may be aggregated into crystalline and paracrystalline areas. The cellulose chains are embedded in an amorphous matrix consisting of hemicelluloses, which are the heteropolymers of galactose, mannose, arabinose, and xylose, and pectins, which are heteropolymers of glucuronic and galacturonic acid which are more or less methoxylated.[56,57,58,59] An attractive model for the primary cell wall has been devised on the basis of selective enzymatic digestion.[60] In this model, xyloglucan is linked to pectic polymers by covalent bonds and to cellulose polymers by hydrogen bonds. The pectic polymers are covalently linked to wall proteins. Thus a mesh of covalently linked pectins, xyloglucans, and protein cross link the cellulose fibers by hydrogen bonding to them. In addition to carbohydrate, the secondary cell wall contains an aromatic polymer, lignin, which after cellulose is the second most abundant organic compound on earth. As a general rule, lignin is present at about 60% of the dry weight level of cellulose and may account for 24% of the total dry weight of wood. There is evidence that lipid and protein may also be an integral part of cell wall.

Lignin merits a detailed consideration. It is a polyaromatic polymer and is rather resistant to chemical degradation. In paper manufacture lignin is removed from wood chips by boiling in alkaline bisulfite to break up the lignin polymer net — giving the vile smell of H_2S characteristic of pulp mills, tons of soluble lignin breakdown products that usually pollute rivers, and the insoluble cellulose used for paper.

For chemical study, lignin may be isolated from finely ground wood that is first extracted with acetone to remove the terpenes. Next, the lignin is extracted in a mixture of dioxane-water to give a soluble, carbohydrate-free material that is very heterogeneous in size and shape. Lignin is relatively resistant to enzymatic degradation and is thus responsible for much of the organic matter in soil. Lignin is also prepared through the activities of "black rot fungi" that digest the cellulose and other carbohydrates leaving lignin. There are less abundant "white rot fungi" that digest the lignin and leave the cellulose.

The structure of lignin has been deduced from incorporation of labelled precursors, the isolation of intermediates, the mechanism of polymerization, and from gross chemical analysis of the product.[61] The specific details are very well reviewed by Brown.[62] One needs only review the biosynthetic scheme outlined in Figure 5.11 and acknowledge that it rests on sound evidence. The aromatic amino acids are synthesized by conventional pathways which are the same in higher plants as in *E. coli* (n.b. lignin is found only in higher plants and not in mosses, algae, or fungi). Although there are several sites of regulation in this sequence, the phenylalanine ammonium lyase is a particularly interesting one since this enzyme is under photocontrol.[63]

Para hydroxy cinnamyl, sinapyl, and coniferyl alcohols are the immediate precursors of lignin. At some stage during their biogenesis it is likely that the plant glucosylates the para hydroxy position to protect these aromatic compounds from oxidation — possibly as early as the coumaric acid step. The glucosides of these three aromatic alcohols are present in woody tissue and are resistant to phenolase attack. A beta glucosidase is found between the cambium and mature wood and this enzyme splits off the sugar and exposes the aglycones, i.e., the aromatic alcohols, to oxidation which in turn leads to their polymerization.

Freudenberg found that incubation of these aromatic alcohols with laccase or peroxidase — the latter enzyme is ubiquitous in plants — leads to formation of lignin-like polymers. He mixed 14 mole% coumaryl alcohol, 80 mole% coniferyl alcohol, and 6 mole% sinapyl alcohol with laccase and got something indistinguishable from spruce lignin. The plant species character of the lignin reflects the relative abundance of each alcohol and in general the hardwoods have a higher methoxyl content in their lignins. By running the polymerization at low pH, it is possible to isolate dimeric intermediates that cannot be found in the intact plant since the reaction is a free radical one. The structure of the

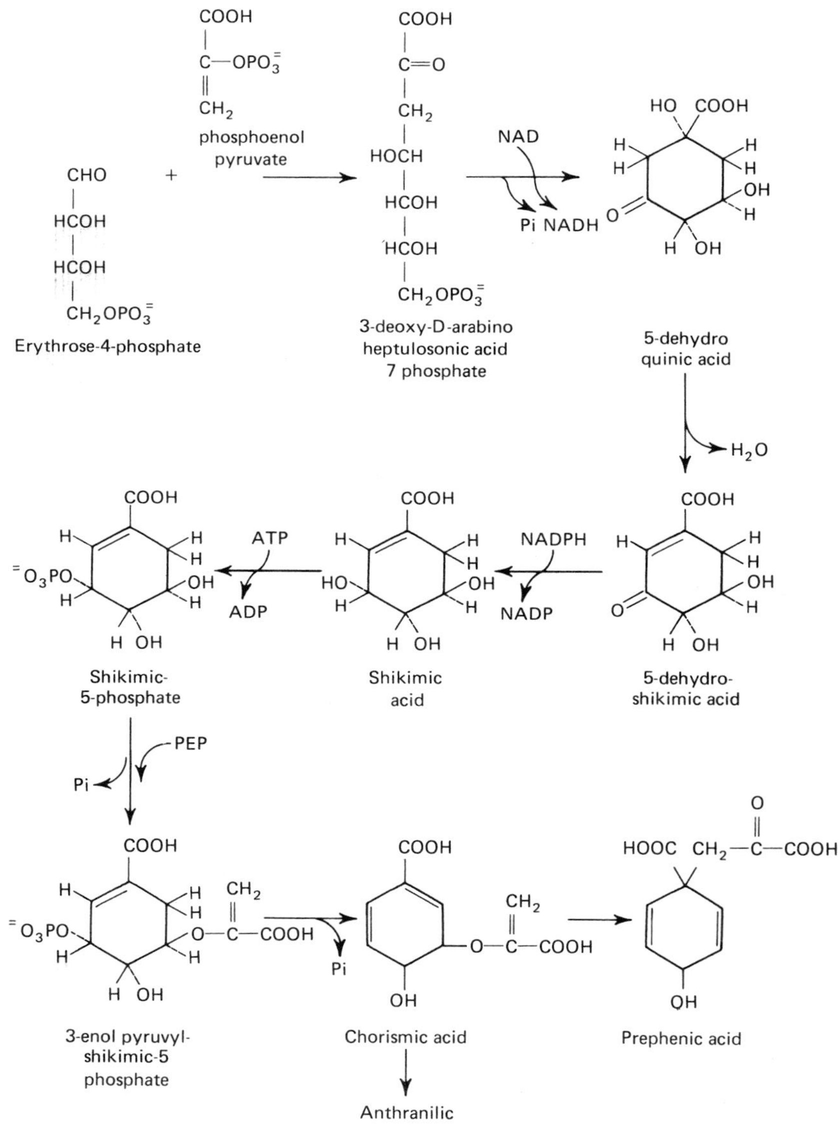

Figure 5.11a   Biosynthesis of lignin precursors.

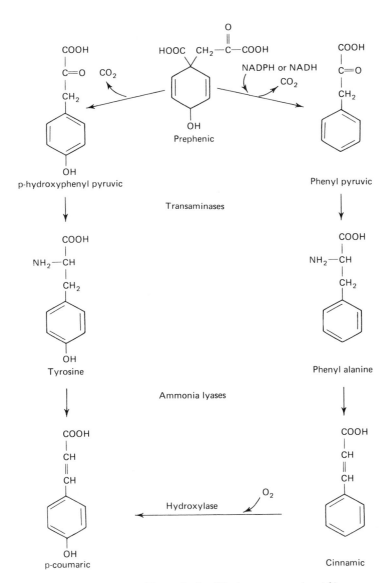

Figure 5.11b Biosynthesis of lignin precursors (*contd.*).

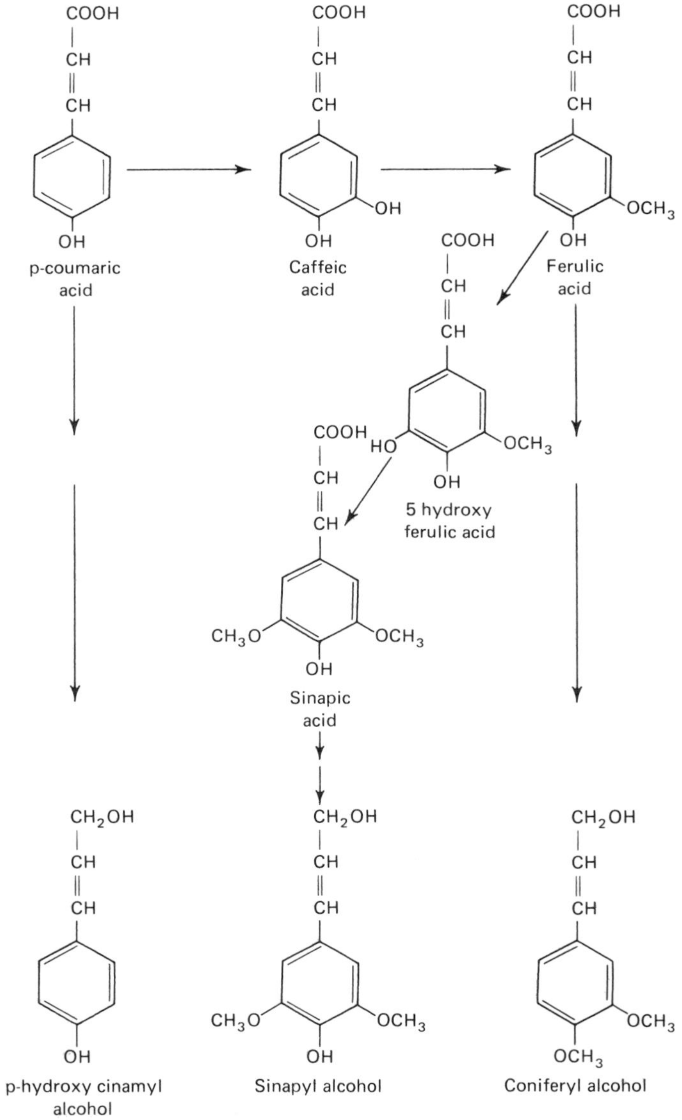

Figure 5.11c   Biosynthesis of lignin precursors (*contd.*).

intermediates suggests the mechanism shown in Figure 5.12. Several other dimers – guiacylglycerol beta coniferyl alcohol, pinoresinol, and bis dehydro-diconiferyl alcohol – were found as intermediates in the synthesis as were

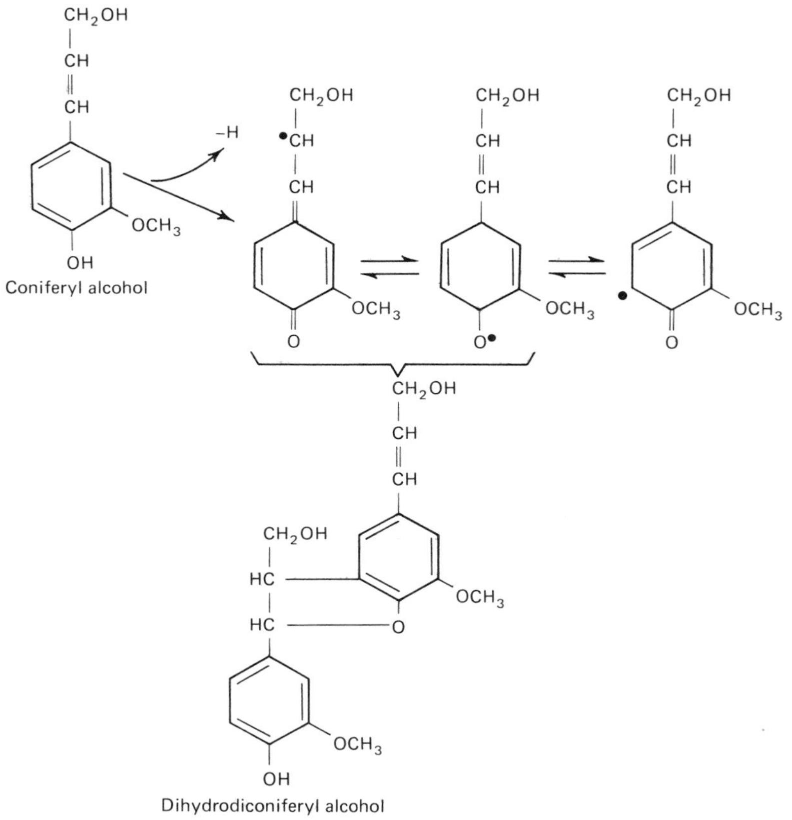

Figure 5.12   The dimerization of lignin precursors by a free radical mechanism.

trimers and tetramers of the aromatic monomer unit. The structures of these intermediates led Freudenberg to suggest the structure for lignin shown in Figure 5.13. It is worth noting that lignin is without optical activity, indicating that optically active precursors are polymerized in a random way. There are no intermediate dimers isolated from chemical degradation, indicating that the polymer is a three-dimensional net. There is no catabolism of lignin.

5.8   CELL WALL PROTEIN

Cell wall protein is a relatively new area of investigation. The cell walls may be isolated from a plant homogenate, washed thoroughly, and checked by electron microscopy for visible cytoplasmic contaminants. Up to 40% of the total

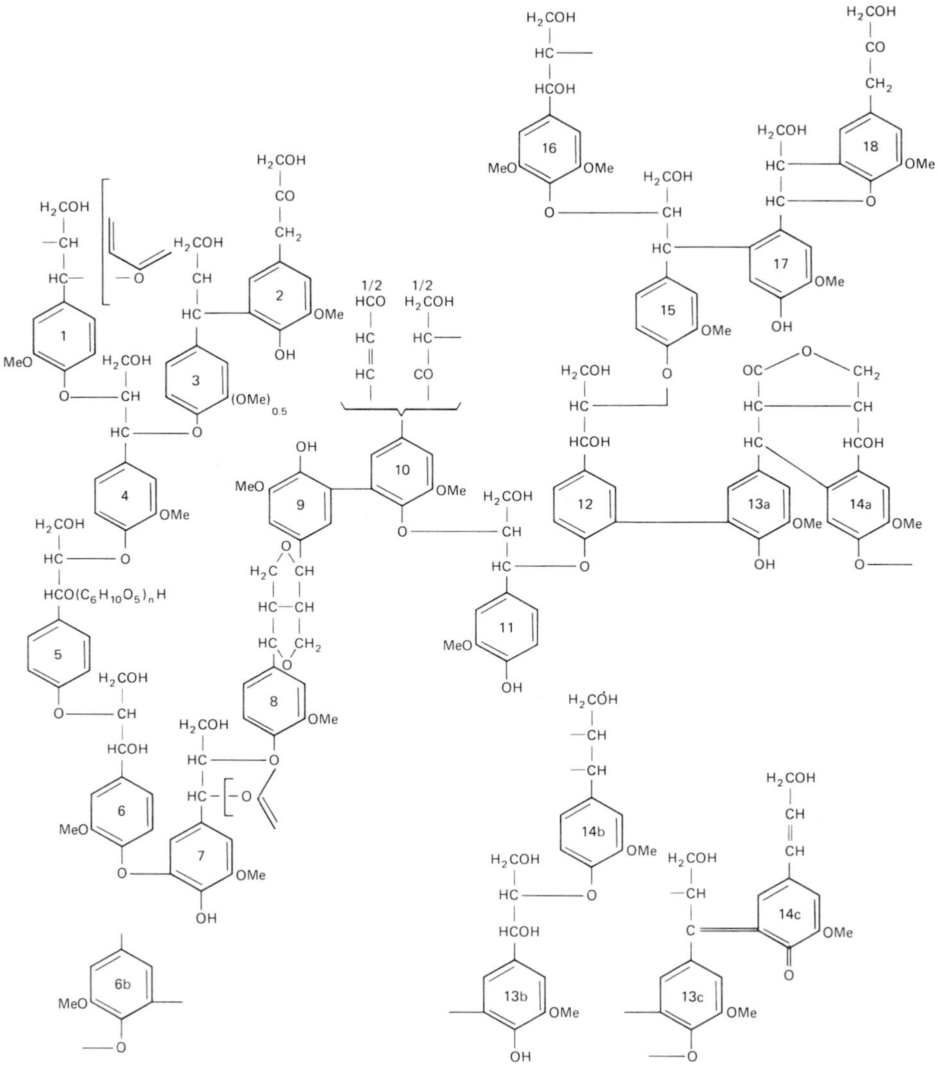

Figure 5.13   Freudenberg's diagrammatic representation of spruce lignin.
[From Freudenberg, K., *Science,* **148**, 595 (1965).]

nitrogen of cultured plant callus tissue remains associated with the insoluble wall fraction. An abundant amino acid in the cell wall protein is hydroxy-proline – 13% of the total – which is similar to the level of that amino acid in collagen which is a major structural protein in animal tissue. A hydroxyproline rich protein is associated with the plant cell wall inner surface where it may be deposited by apposition.[64] By chemical analysis of wall fractions from cultured sycamore cells, Lamport found that 90% of the cell hydroxyproline is in the wall protein.[50] Since hydroxyproline analogs are good inhibitors of cell growth, Lamport suggests that the hydroxyproline containing protein is a controlling factor in plant growth, perhaps by forming labile cross links with the cellulose microfibrils. Since there are several reports on the isolation of hydroxyproline rich mucopolysaccharide or glycoprotein, the chemical details of these polymers should be forthcoming.[65,66] Cleland has found hydroxyproline in three fractions of *Avena* – 60% is bound to the cell wall and 40% is in the cytoplasm. Of the cytoplasmic fraction, $\frac{1}{2}$ is trichloroacetic acid insoluble therefore in protein and $\frac{1}{2}$ is trichloroacetic acid soluble but is not dialyzable.[67] This latter fraction is assumed to be hydroxyproline bound to a polysaccharide and might be a wall precursor, although it is hard to imagine how an amino acid with a big sugar polymer attached can get into a protein. Furthermore, hydroxyproline in protein arises by hydroxylation of proline in the protein, not via incorporation of free hydroxyproline. Crispeels has found evidence for hydroxylation of proline in cytoplasmic protein which might then be deposited in the wall.[68,69]

Holleman found direct incorporation of hydroxyproline into protein of sycamore cells incubated at growth-inhibiting levels of hydroxyproline.[70] Alpha-alpha dipyridyl, an iron chelator, was used to suppress proline hydroxylation. In this situation, it appeared that hydroxyproline inhibited growth by getting into proteins where proline belonged. Cleland used the unnatural cis hydroxyproline to refine this experiment because this isomer cannot be converted to proline.[71] He found that most of the cis hydroxyproline is incorporated into cytoplasmic protein. Therefore, the inhibition of growth by high concentrations of hydroxyproline probably has nothing to do with cell walls. Cleland had found that hydroxyproline at non-inhibitory levels goes into protein only after conversion to proline.[72] Finally, Cleland found hydroxyproline proteins increased markedly in the cell wall of pea during the transition from elongation to stationary size and concluded that the hydroxyproline protein may stop elongation.[73]

The synthesis of cell wall protein poses some interesting problems. Although it appears that the hydroxylation of proline occurs at the polypeptide level, one needs more information about the formation of protein polysaccharide bonds. The incorporation of glucosamine into cell wall glycoproteins has been observed.[74] The isolation of ribosomes from cell wall preparations of barley suggests that these ribosomes were incorporated into the primary wall structure during its formation.[75] Must the wall protein be delivered to the site of apposition on the wall by the ribosome? Does the covalent attachment to polysaccharide occur while the protein is still attached to the ribosome?

5.9   THE PLASMA MEMBRANE

Beneath the wall is a cell membrane and its chemistry may be approachable. Thimann prepared protoplasts from *Avena* coleoptile by digesting the cell wall with mixed carbohydrase enzymes from the snail.[76] The protoplasts were osmotically sensitive but in isotonic medium these protoplasts did not burst when treated with lipase and protease. With the wall gone, one would think the membrane could be attacked by these hydrolytic enzymes. Treatment of these protoplasts with detergents and polyene antibiotics leads to the inference that there is little sterol in the membrane; that would be an important departure from the pattern of membrane construction thought to be general in eucaryotic cells.[77] A variety of plant species can be used to make protoplast.[78] In one interesting case, tomato cotyledon cells grown in culture were converted to protoplasts.[79] Given enough time, the protoplasts will regenerate their cell walls, but if one is prompt, the freshly prepared protoplasts will show pinocytosis of ferritin molecules.[80] Cell fusion of protoplasts from tomato with protoplasts from tobacco has been achieved and this opens enormous possibilities for genetic engineering of plants.[81]

GENERAL REFERENCES

Cocking, E. C. "Plant Cell Protoplasts — Isolation and Development," *Ann. Rev. Plant Physiol.*, **23**, 29 (1972).

Hassid, W. Z. "Transformations of Sugars in Plants," *Ann. Rev. Plant Physiol.*, **18**, 253 (1967).

Lamport, D. T. A. "Cell Wall Metabolism," *Ann. Rev. Plant Physiol.*, **21**, 235 (1970).

Loewus, F. "Carbohydrate Interconversions," *Ann Rev. Plant Physiol.*, **22**, 337 (1971).

Northcote, D. H. "Chemistry of Plant Cell Walls," *Ann Rev. Plant Physiol.*, **23**, 113 (1972).

6 ‖ LIPIDS AND LIPID PIGMENTS

Lipids and their metabolism in plant material are not very different from that in animals and bacferia.[1,2,3] Fatty acid synthesis, desaturation, and incorporation into plant structures show some interesting regulatory and evolutionary features. The special mechanisms for synthesis of wax, cutin, chlorophyll, and carotenoids are unique and of great significance in the texture and color of the world.

6.1 BIOSYNTHESIS OF LIPIDS

Fatty acid synthesis follows the usual pattern, as outlined in Figure 6.1. The following special points bear noting:

1. Higher plant material has ACP (acetyl carrier protein) such as is found in some bacteria.[4] Matsumura and Stumpf have purified the ACPs from spinach and avocado and have published a partial amino acid

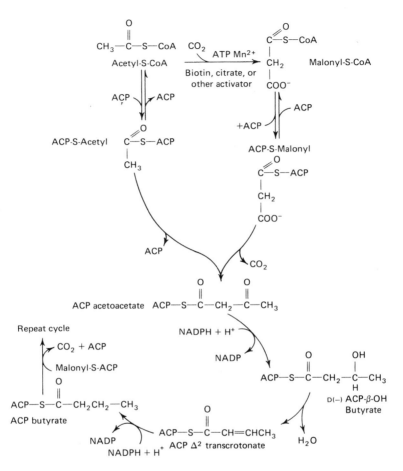

Figure 6.1   Pathway of fatty acid synthesis. Note the importance of ACP and of the carboxylation and decarboxylation as parts of the reaction sequence.

sequence.[5] The spinach protein is very similar to *E. coli* ACP – mol. wt. 10,000 – and has the 4'phosphopantotheine attached via its phosphate to a serine in the protein (Fig. 6.2).

2. Malonyl CoA comes from the standard acetyl CoA carboxylase and this enzyme is avidin sensitive, therefore biotin dependent, in both mesocarp and chloroplast preparations of the enzyme.

$$HCO_3^- + ATP + \text{biotin-enzyme} \rightleftharpoons {}^-OOC\text{-biotin-enzyme} + ADP + P_i$$

$${}^-OOC\text{-biotin-enzyme} + \text{acetyl-S-CoA} \rightleftharpoons \text{malonyl-S-CoA} + \text{biotin-enzyme}$$

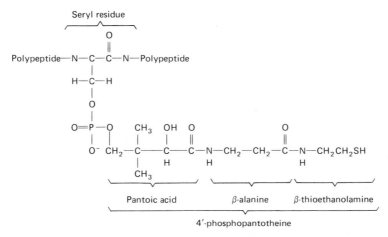

Figure 6.2   The active site region of the acetyl carrier protein which attaches through the sulfhydryl group on the beta thioethanolamine, the acetyl or fatty acyl residue.

However, the alga *Euglena* yields an enzyme preparation that synthesizes fatty acid in an avidin insensitive, bicarbonate insensitive sequence of reactions that prefers acetyl CoA to malonyl CoA as substrate.

3. Fatty acid synthesis in animals and yeast is catalyzed by a multienzyme complex presumably localized in the soluble fraction of the cell. The system from soybean and potato is soluble; but in avocado, it is found on the mitochondria.[6,7] In green leaves, fatty acid synthetase activity is found in the chloroplasts.

Euglena gracilis has two distinct systems for synthesizing fatty acids.[8] When the cells are grown heterotrophically in the dark, they contain a high molecular weight fatty acid synthetase that does not require any added ACP and resembles the fatty acid synthetase found in yeast and in mammalian tissues. If *Euglena* is grown autotrophically in the light, the above fatty acid synthetase is retained and an additional one appears. The light induced synthetase is not a multienzyme aggregate and does require ACP. This suggests that a detailed study of higher plant fatty acid synthetases is needed to see if the cytoplasmic enzymes resemble those of yeast and mammals. The higher plant chloroplast fatty acid synthetases show a similarity to the bacterial enzymes. Chloramphenicol but not cycloheximide prevents the appearance of the light-induced fatty acid synthetase which accompanies greening and which presumably resides in the chloroplasts of *Euglena*.[9] This indicates that the new enzyme is synthesized on chloroplast ribosomes rather than on cytoplasmic ribosomes. Since the chloroplast ribosomes are similar to those found in procaryotes, it is appropriate that the procaryotic type fatty acid synthetase be assembled on them. It remains to be seen if chloroplast DNA codes this enzyme system and if similar distinctions exist between chloroplast and cytoplasmic fatty acid synthetases of higher plants. In any case, the diversion of newly fixed carbon from carbohydrate to fat in the chloroplast is not enormous and probably is a reflection of the chloroplast's ability to synthesize itself rather than to serve as a fat depot.

The fatty acid synthesizing systems can run out long chain saturated fatty acids, and specific control points may regulate the length of the products.[10] However, the major fatty acids in leaves are unsaturated and the chloroplasts have a high proportion of polyenoic fatty acids compared to the rest of the cell. In the chloroplast, ATP levels may regulate the extent of desaturation.[11] For a time it was thought that blue-green algae were exceptional in lacking polyunsaturated fatty acids but examination of more species showed that only *Anacystis nidulans* lacks polyenoics, but other species have plenty.

The introduction of double bonds is an oxygen dependent desaturase mechanism in higher plants and has been described by tracer experiments with cell-free preparations but not with pure enzymes. Bloch's work on *Euglena* is probably the most detailed.[12,13] Bleached *Euglena* has a particulate acyl-CoA desaturase like yeast, involving oxygen and NADPH in a hydroxylation followed by dehydration sequence. In the green photoautotrophically grown cells, one finds fatty acid synthesis proceeding on an ACP sequence as in the bacteria and the desaturation involves three proteins.

$$h\upsilon \searrow$$

NADPH $\longrightarrow$ flavoprotein $\longrightarrow$ ferredoxin $\longrightarrow$ desaturase

Here, reduced ferredoxin is the electron donor in the oxygenation catalyzed by the desaturase. Reduced ferredoxin could arise either by direct chloroplast reduction or by getting electrons from NADPH via the NADPH-ferredoxin oxidoreductase. The significance of the parallel appearance of unsaturated fatty acids and Hill reaction activity in greening *Euglena* is best rationalized as the result of the appearance of ferredoxin. Light-induced greening is accompanied by the appearance of ferredoxin that allows the ferredoxin-dependent desaturation of fatty acids.

The major fatty acids in leaves are listed in Table 6.1.

TABLE 6.1   THE MAJOR FATTY ACIDS IN LEAVES

Palmitic acid	C_{16}	hexadecanoic acid (slight)
Palmitoleic	C_{16}	$\Delta 9$ hexadecenoic acid
Stearic	C_{18}	octadecanoic acid (slight)
Oleic	C_{18}	$\Delta 9$ octadecenoic acid
Linoleic	C_{18}	$\Delta 9,12$ octadecadienoic acid
Linolenic	C_{18}	$\Delta 9,12,15$ octadecatrienoic acid

Direct conversion of stearic to oleic has been shown with plant material. Desaturation of oleic is catalyzed by a microsomal enzyme in maturing safflower seed and oleyl CoA is the substrate for this reaction.[14] In *Chlorella* and pumpkin leaves, the conversion of oleic to linoleic proceeds as the phosphatidyl-choline derivative.[15,16]

Ricinoleic acid is 12 hydroxy oleic acid found in the castor bean and is formed via

$$\text{Oleyl CoA} \xrightarrow[\text{NADH}]{\text{oxygen}} \text{ricinoleyl CoA}$$

— a hydroxylation rather than by oxidation of oleic to linoleic than hydration of the double bond.[17,18]

There are cyclopropane fatty acids in hibiscus leaf and the wonderful cyclopentane fatty acids from Chaulmoogra nut oil, but never mind.

Phosphatidic acid is synthesized in plants along conventional lines.

Glycerol phosphate + fatty acyl-CoA

Lysophosphatidic acid + fatty acyl-CoA

Phosphatidic acid

The next step is usually to dephosphorylate this phosphatidic acid and react the alpha, beta diglyceride with a third fatty acyl-CoA to give the triglyceride. In leaves, however, the major lipid is not triglyceride but a galactosyl glyceride. Chloroplasts are especially active in incorporating C^{14} galactose from UDP-galactose into galactolipid[19] (Fig. 6.3). Related to the galactosyl diglyceride in

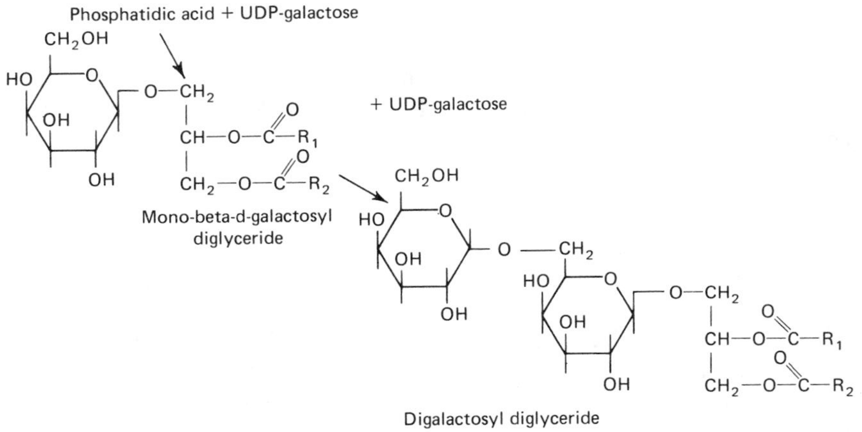

Figure 6.3   The biosynthesis of digalactosyldiglyceride.

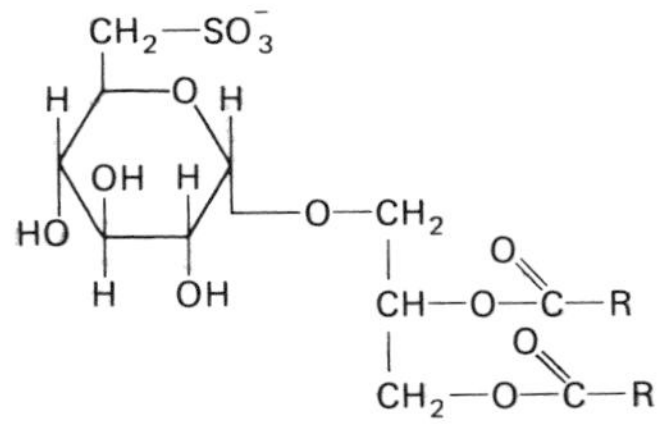

Figure 6.4   Sulfoquinovosyl diglyceride.

appearance and subcellular localization is the sulfoquinivosyl diglyceride (Fig. 6.4). The origin of sulfoquinivose is not clear (note that this is a C–S bond and not the usual C–O–S sulfate ester of a sugar). One might imagine phosphoadenosine phosphosulfate (PAPS) sulfurylation of quinivose or of a nucleotide of this sugar. Alternately, sulfoquinivose might be formed by

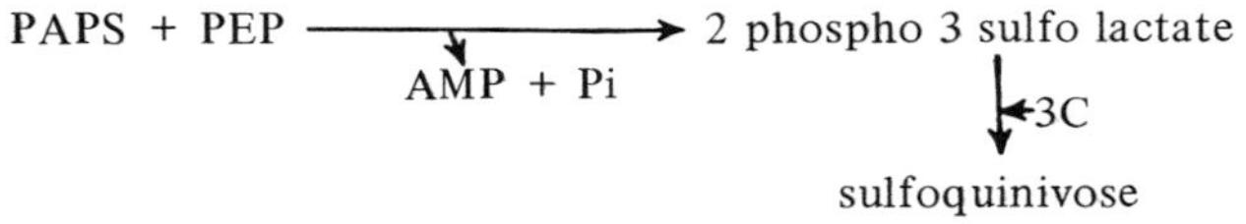

$$\text{PAPS + PEP} \xrightarrow[\text{AMP + Pi}]{} \text{2 phospho 3 sulfo lactate} \xrightarrow{\text{3C}} \text{sulfoquinivose}$$

Linolenic acid is the chief fatty acid in these glycerides, especially in the chloroplast. The chloroplast has little triglyceride and little phospholipid. Most of the lecithin is in the mitochondrial and microsomal fractions.

Plant synthesis of phospholipid is summarized by the diagram in Figure 6.5 and is quite similar to the patterns found in animal tissue. The phospho, glyco, and sulfolipids are mainly associated with membranes in which they are found with protein. In the higher plants these membranes contain sterols such as sitosterol and spinasterol.

Many plant cells contain osmiophilic globules which are collections of a variety of lipids – galactolipids, triglycerides, carotenoids, etc. – which may represent mere coalescing of extra lipophilic substances in the cell. In addition, some plant tissues contain oleosomes that are bounded by a single membrane and that contain lipid and little else.[20]

6.2   CUTIN AND WAX

The surface lipids of plants represent a structural lipid in which carbon is irreversibly invested to provide a water repellent surface that both protects the plant from external assault and conserves water within.[21,22] This cuticle layer determines the effectiveness of foliar applicants and possibly provides a barrier to infection. A representative cuticle analysis shows waxes – esters of fatty acid and fatty alcohol C_{10} to C_{30}, 15%; saponin glycosides, 10%; cellulose, 14%; and

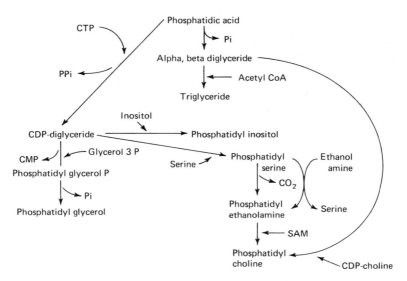

Figure 6.5 Summary diagram of phospholipid synthesis.

cutin – a hydrocarbon net of straight chain paraffins of C_{21} to C_{35}, 61%. The cuticle layer in tropical plant leaves is generally heavier and contains more saturated, hence higher melting, wax than those plants in temperate climates.

The waxes are formed from long chain fatty acids, part of which are reduced to alcohols. Cabbage shows three separate enzymes that participate in wax synthesis.

$$R - C \overset{O}{\underset{S-CoA}{\big<}} + R'OH \longrightarrow R - C \overset{O}{\underset{OR'}{\big<}} + CoA\ SH$$

$$R - C \overset{O}{\underset{phospholipid}{\big<}} + R'OH \longrightarrow R - C \overset{O}{\underset{OR'}{\big<}} + \text{deacylated phospholipid}$$

$$R - C \overset{O}{\underset{OH}{\big<}} + R'OH \longrightarrow R - C \overset{O}{\underset{OR'}{\big<}} + H_2O$$

The last of these reactions is a simple esterase activity with a favorable equilibrium.

The paraffins represent a more intriguing problem since they are odd-numbered carbon chains. The usual acetate to C_{18} fatty acid provides a start. Then continued C_2 addition by what may be a distinct path gives something as long as a C_{36} which must be decarboxylated to give the hydrocarbon of odd-number chain length.[23]

The cutin also contains cross-linked acids and alcohols (Fig. 6.6), and cell-free reactions for hydroxylation and epoxidation of the precursors are known.[24]

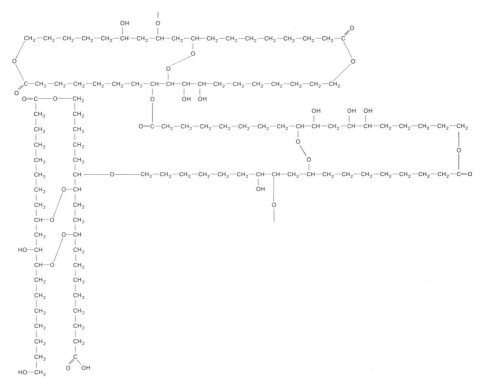

Figure 6.6   Repeating units in cutin.

6.3   LIPID DEGRADATION

Lipid degradation in plant materials is well documented and shows some novelties. There is an enormous literature on enzymes that hydrolyze galactolipids, phospholipids, and triglycerides in plant tissue. The lipases that hydrolyze triglycerides to glycerol and fatty acid make physiological sense in germinating seeds, where they liberate reserve carbon for metabolic use.[25] In the ungerminated castor bean, lipases are localized in oleosomes which are the fat storage organelles mentioned earlier.[26] One wonders if these lipases might be

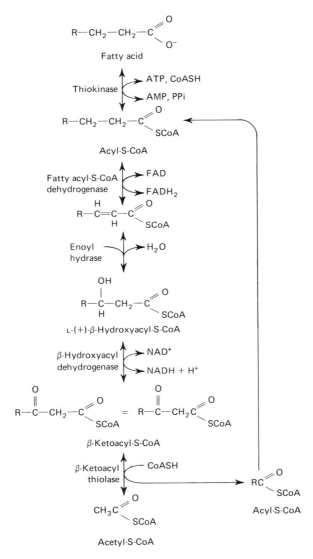

Figure 6.7   The beta oxidation sequence for the conversion of fatty acids into acetyl CoA.

activated by hydration, for there is no obvious need for proteolytic activation of an "ogen" precursor or for *de novo* synthesis as seen with carbohydrate releasing enzymes of germinating seeds.

The fatty acids so liberated are usually converted by the conventional beta oxidation sequence to acetyl CoA as shown in Figure 6.7. Not all of these enzymes have been isolated from plant tissues, but transit of label through the intermediates is convincing evidence of the sequence. The beta oxidation sequence in germinating castor bean is now clearly localized in the glyoxysome.[27] Thus the glyoxysome contains not only the glyoxylate cycle enzymes to convert acetate into a carbohydrate precursor but also the enzymes for the conversion of fatty acid into acetate.

Stumpf has presented evidence for a second beta oxidation system in the mitochondria of some plant tissue where acetyl CoA would be pumped into the Krebs cycle for energy generation. Since fatty acids enter mammalian mitochondria as the carnitine ester and since carnitine is available in plants, it will be interesting to see if an acyl-CoA-carnitine transferase activity is available to catalyze a similar mitochondrial entry mechanism in plants.[28]

The oxidation of unsaturated fatty acids in plants involves some special problems in that the oxidation mechanism of desaturation gives a cis double bond while the flavoprotein of beta oxidation gives a trans double bond.

$$\gamma \quad \beta \quad \alpha$$
$$C-C-C-C-C-C=C-C-C=C-C-C|C-C|C-C|C-COOH$$

When linoleic acid is degraded, it will be disassembled to a beta-gamma cis system. There is an isomerase that will convert this to an alpha-beta trans for conventional hydration.

$$\beta \quad \alpha$$
$$C-C-C-C-C-C=C-C|C-C|C-C|C-C|C-C|C-COOH$$

Once past the first double bond, the second is an alpha-beta cis double bond which is hydrated to the D (–) beta hydroxy acyl-CoA which is epimerized to the usual L (+) isomer and this can be oxidized by the beta hydroxyacyl-CoA dehydrogenase. Thus the unsaturated fatty acids are converted to acetyl CoA. The pathway for ricinoleic acid degradation in germinating castor bean has been worked out and is described by the accompanying diagram in Figure 6.8.[29]

An alternative to the beta oxidation pathway of fatty acid breakdown has been revealed by the studies of Stumpf – the alpha oxidation cycle. This sequence is inferred from tracer studies and isolation of labelled intermediates rather than the isolation of individual enzymes. The activity is associated with a particulate fraction from peanut cotyledon.

1. $R - CH_2 - CH_2 - COOH + H_2O_2 \longrightarrow CO_2 + R - CH_2 - CHO$

2. $R - CH_2 - CHO + NAD \xrightarrow{\text{aldehyde dehydrogenase}} R - CH_2 - COOH + NADH$

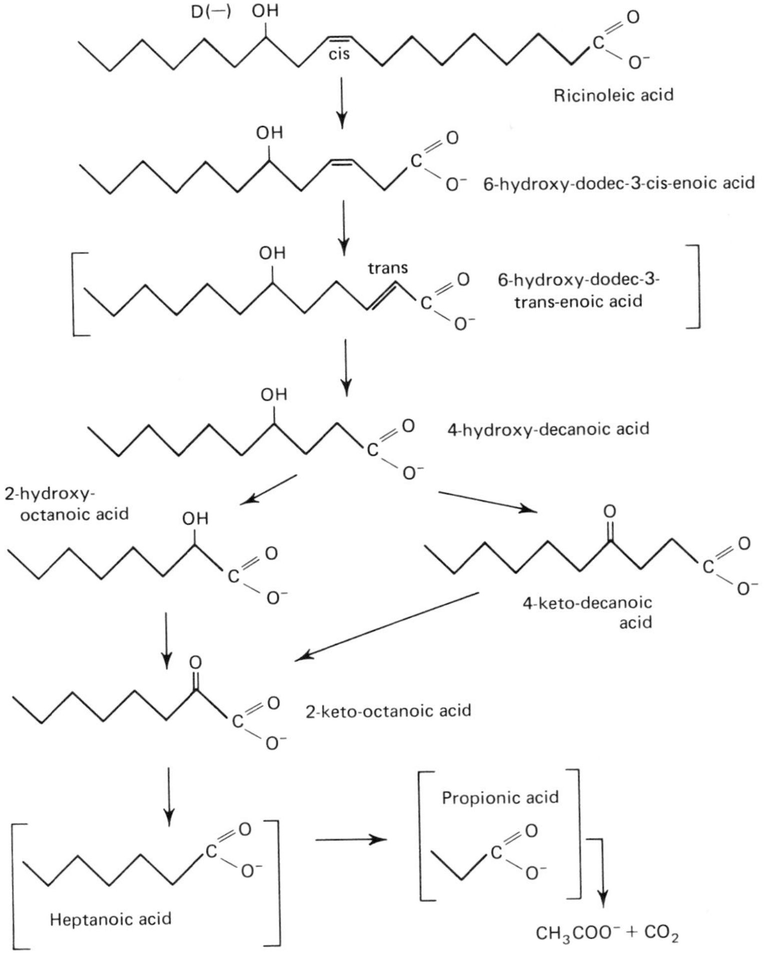

Figure 6.8   A proposed pathway for ricinoleic acid catabolism in the germinating castor bean. [From Hutton, D. and P. K. Stumpf, *Arch. Biochem. Biophys.*, **142**, 48 (1971).]

Hydrogen peroxide to drive the first reaction might be generated as a product of the glycolate oxidase and the aldehyde dehydrogenase is available in the tissue. The real question is the purpose that this pathway might serve. Only longer fatty acids are susceptible to alpha oxidation leaving a moderately long chain residue that must be sent through the beta oxidation scheme. Although the alpha oxidation sequence might repeat several times on the same fatty acid, it could result in an odd-numbered chain length that on beta oxidation would finally yield propionate as the CoA ester. The propionyl CoA could then be used in one of the reactions shown in Figure 6.9.

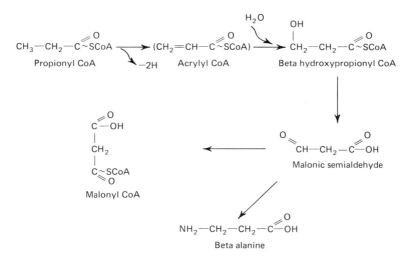

Figure 6.9   The utilization of propionyl CoA.

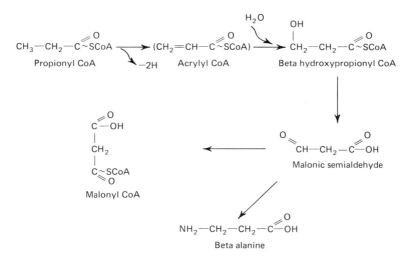

Figure 6.10   An alpha oxidation pathway initiated by an oxygenation reaction.

A recent variation of the alpha oxidation path has been discovered in young, green leaf tissue by James.[30] In this system oxidation proceeds not via an aldehyde but via an alpha hydroxy acid which is then decarboxylated as shown in Figure 6.10. The reactions are postulated from tracer studies through the intermediates. If the reductant for the alpha hydroxylase reaction is a high redox potential compound like ascorbate rather than NADH, this sequence might be more efficient at energy conservation. The amount of carbon going through either of these alpha oxidation sequences, however, appears small and their physiological role is still a matter of speculation.

6.4   ISOPENTENE UNITS AND TERPENES

The prevalence of terpenoids in plants demands a brief look at this diverse and abundant class of compounds including carotenoids, oils, sterols, pigments, terpenes, glycosides, and many other related compounds. Note especially that terpenes contribute to the structure of the phytol chain on chlorophyll, the phytosterols, kaurine – the precursor of gibberellins, the side chains of plastoquinone, CoQ, tocopherols, vitamin K, and to rubber. Robinson's book is an excellent wedge into this and other types of "secondary" plant products – compounds that are off the main line of metabolism.[31] Among the terpenes, the basic building block is the isopentane unit, and, with a few notable exceptions, multiples of this unit are most commonly linked head to tail.

Tradition has designated the monoterpenoid as C_{10} with two isopentane units. Geranyl pyrophosphate is probably the most familiar example. The diterpenoid (C_{20}) geranylgeranyl pyrophosphate dimerizes head to head to give the C_{40} carotenoids just as the sesquiterpenoid C_{15} farnesyl pyrophosphate dimerizes head to head to give squalene and on to the sterols.

The biogenesis of the C_5 unit has been worked out with yeast and pigeon liver systems (Fig. 6.11) and the same reactions are presumed to occur in plants in which a few of the intermediates and some tracer evidence supports the pathway similar to that in yeast.[32,33] Good cell-free preparations for this sequence are hard to obtain.

6.5   CAROTENE BIOSYNTHESIS

The stepwise $C_5 \longrightarrow C_{10} + C_{10} \longrightarrow C_{20}$ sequence to carotenoids with subsequent head to head condensation to C_{40} has been confirmed with tomato and spinach enzyme preparations[34] (Fig. 6.12).

The C_{40} must be isomerized and put through four successive dehydrogena-

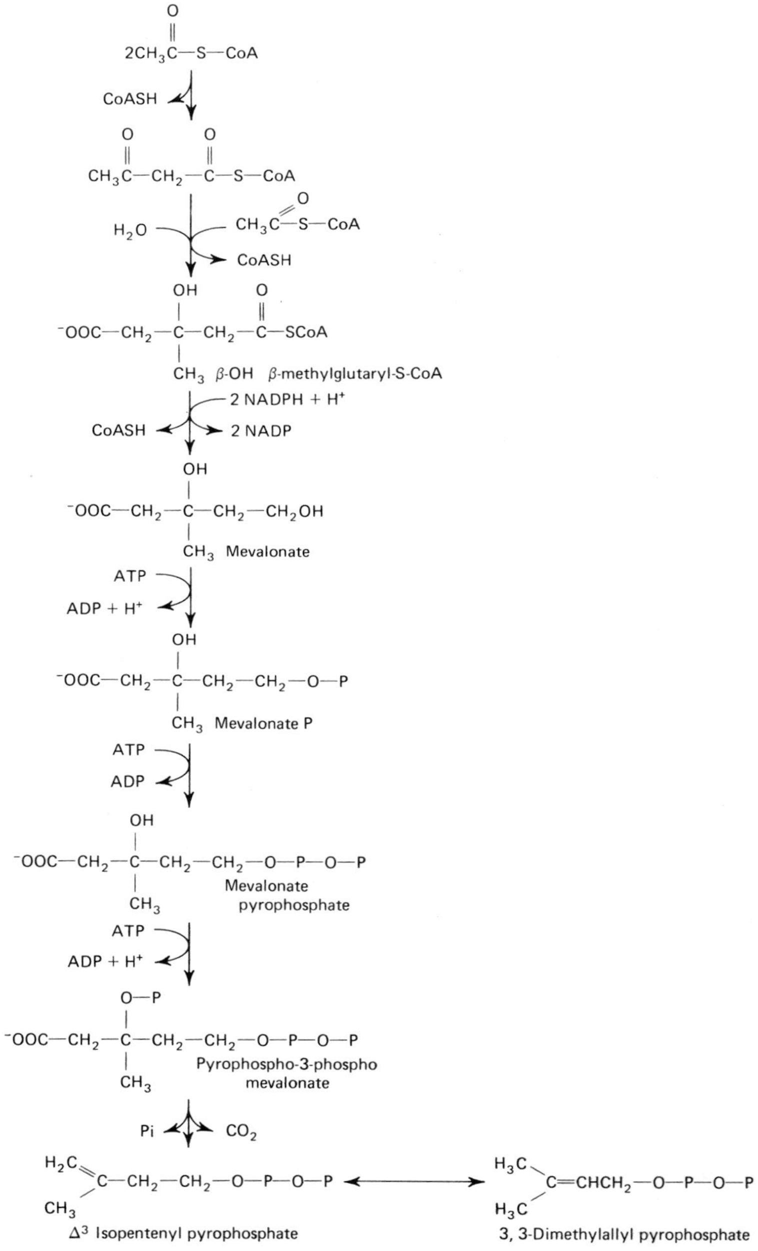

$$2CH_3\overset{\displaystyle O}{\overset{\|}{C}}-S-CoA$$

CoASH

$$CH_3\overset{\displaystyle O}{\overset{\|}{C}}-CH_2-\overset{\displaystyle O}{\overset{\|}{C}}-S-CoA$$

H_2O $CH_3\overset{\displaystyle O}{C}-S-CoA$

CoASH

$$^-OOC-CH_2-\overset{\displaystyle OH}{\underset{\displaystyle CH_3}{C}}-CH_2-\overset{\displaystyle O}{\overset{\|}{C}}-SCoA$$ β-OH β-methylglutaryl-S-CoA

2 NADPH + H$^+$

CoASH   2 NADP

$$^-OOC-CH_2-\overset{\displaystyle OH}{\underset{\displaystyle CH_3}{C}}-CH_2-CH_2OH$$ Mevalonate

ATP

ADP + H$^+$

$$^-OOC-CH_2-\overset{\displaystyle OH}{\underset{\displaystyle CH_3}{C}}-CH_2-CH_2-O-P$$ Mevalonate P

ATP

ADP

$$^-OOC-CH_2-\overset{\displaystyle OH}{\underset{\displaystyle CH_3}{C}}-CH_2-CH_2-O-P-O-P$$ Mevalonate pyrophosphate

ATP

ADP + H$^+$

$$^-OOC-CH_2-\overset{\displaystyle O-P}{\underset{\displaystyle CH_3}{C}}-CH_2-CH_2-O-P-O-P$$ Pyrophospho-3-phospho mevalonate

Pi CO_2

$$\underset{\displaystyle CH_3}{\overset{\displaystyle H_2C}{>}}C=CH_2-CH_2-O-P-O-P$$ Δ^3 Isopentenyl pyrophosphate

⟷

$$\overset{\displaystyle H_3C}{\underset{\displaystyle H_3C}{>}}C=CHCH_2-O-P-O-P$$ 3, 3-Dimethylallyl pyrophosphate

Figure 6.11   The conversion of acetate to the 5-carbon isoprenoid precursor.

94

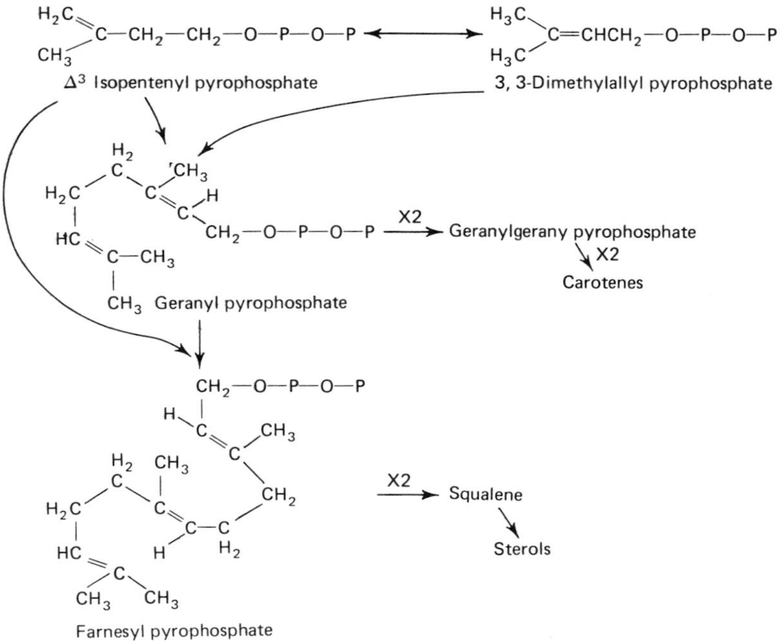

Figure 6.12   Successive dimerization of isoprene subunits to give the carbon skeletons for carotenes and sterols.

tions to give the properly unsaturated chain as outlined in Figure 6.13. This C_{40} unsaturated chain must then by cyclized to give a six-member ring at first one end and then the other. The sequential dehydrogenation and cyclization reactions are known mainly from tomato genetics and *Chlorella* mutant studies. The cyclization mechanism has been elaborated on in cell-free preparations.[35] The final hydroxylation and epoxidation steps are known from O^{18} measurements, and C^{14} S adenosyl methionine is the donor for methoxylation of the carotenoid hydroxyl groups.

Certain herbicides show promise of being selective inhibitors of specific steps in the path of carotenoid biosynthesis.[36] Dichlormate causes accumulation of zeta carotene and amitrole and pyriclor cause a buildup of phytofluene and phytoene as well as zeta carotene. These compounds do not interfere with chlorophyll synthesis but probably cause its photodestruction by preventing the synthesis of carotenoids needed to properly position the chlorophyll or to protect it from photooxidation.

Various isomers of gamma and beta carotene are achieved by shifting the double bond in the ring. The carotene structure can then undergo sequential hydroxylation followed by sequential epoxidation as shown in Figure 6.14.

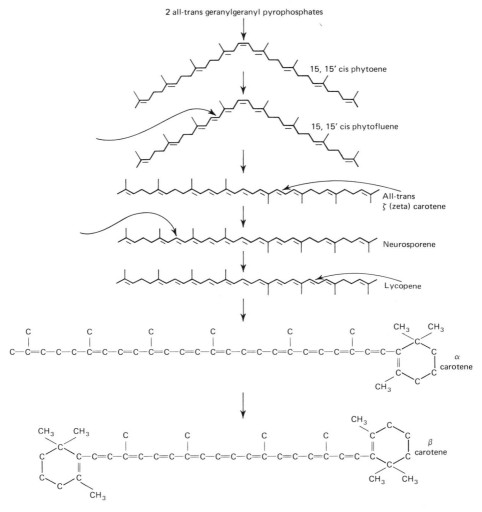

Figure 6.13   The formation of β carotene. [Redrawn from Porter, J. W., and
D. G. Anderson, *Ann. Rev. Plant Physiol.*, **18**, 197 (1967).]

6.6   OTHER TERPENOIDS

Much elegant chemistry has been done in describing the structure, stereochemistry, and biosynthesis of many of the sterols found in plants.[37] Like the mammalian pathway, the sesquiterpene derivative farnesyl pyrophosphate is

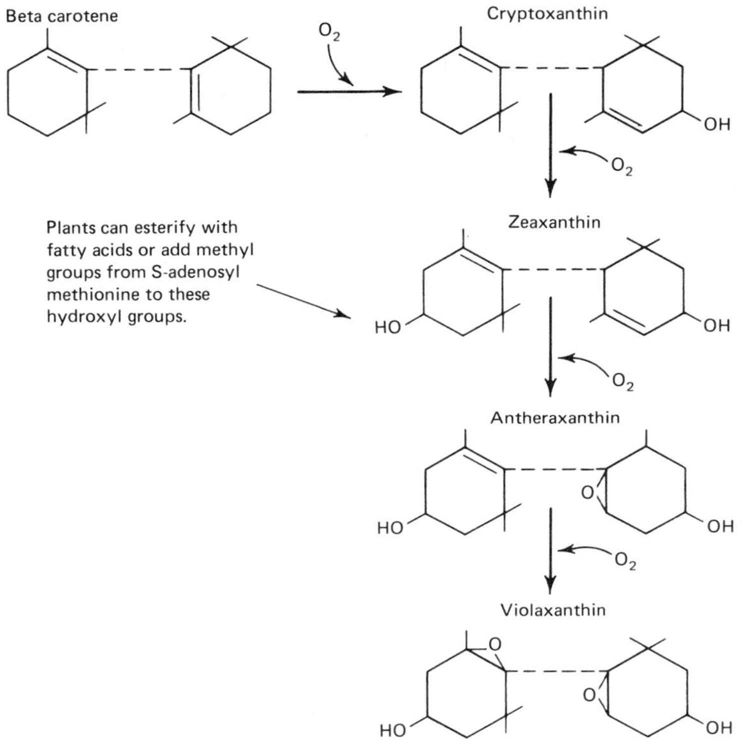

Figure 6.14   Sequential hydroxylation and epoxidation of carotene.

dimerized to give squalene which is then cyclized to provide the sterol ring systems (Fig. 6.15). The diterpene geranylgeranyl pyrophosphate may be cyclized and ultimately converted to a gibberellin[38,39] or may serve as the precursor of the phytol side chain of chlorophyll. The terpenoids also provide the basic biosynthetic unit for rubber formation[40] and in the form of the fossilized plant resins known as amber, they are providing clues to the taxonomy and evolution of ancient plants.[41]

6.7   CHLOROPHYLL SYNTHESIS

The path for porphyrin synthesis was elucidated by the studies of Granick and of Bogorad using plant tissue and by Shemin using animal tissue.[42,43] Subsequently, cell-free systems that synthesize chlorophyll from simple metabolites have been obtained.[44,45]

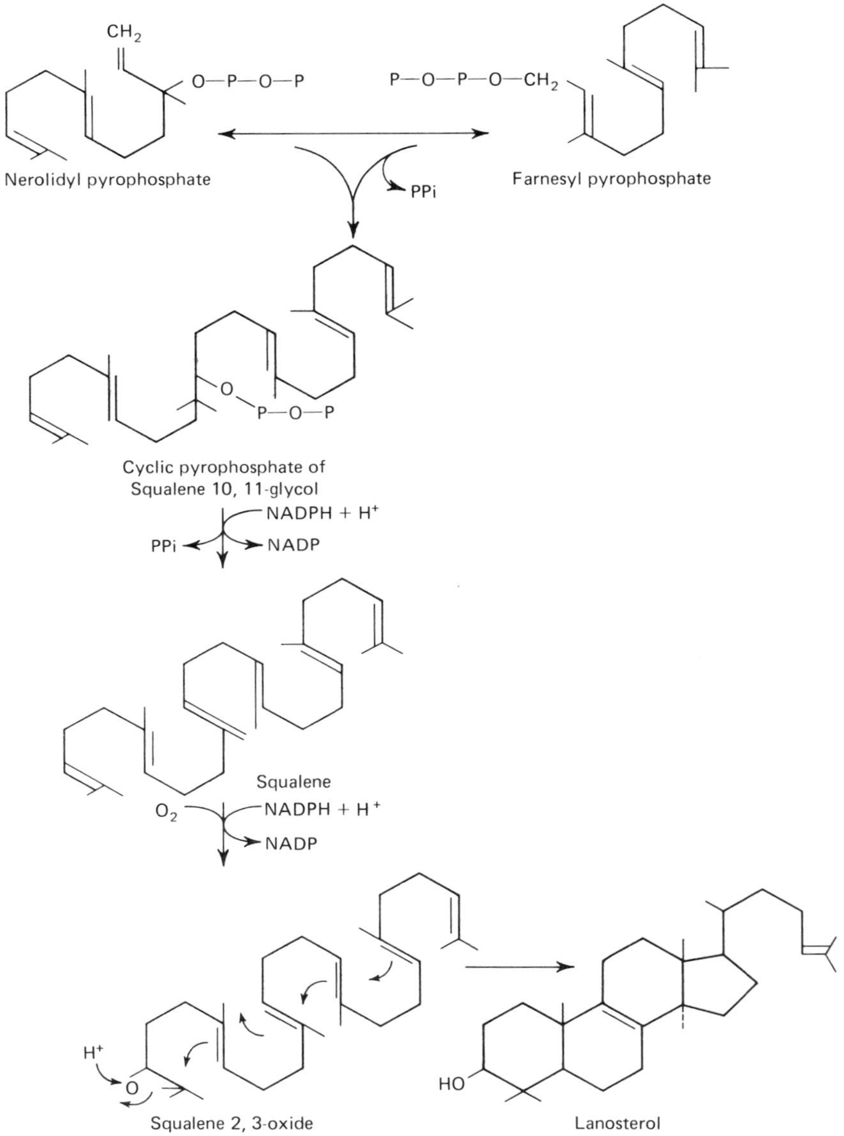

Figure 6.15   The synthesis of the sterol nucleus. Hydroxylation, oxidation, epoxidation, etc. lead to vast numbers of plant sterols.

An outline of the reaction sequence that generates porphyrin rings for chlorophylls, cytochromes, catalase, etc. is as follows:

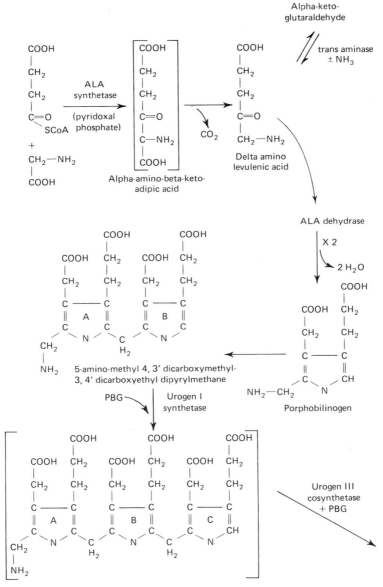

Figure 6.16a   Biogenesis of chlorophyll.

1. Combination of glycine and succinyl CoA to give delta amino levulinic acid (ALA).

2. Combination of two molecules of ALA to give porphobilinogen (PBG) — the single pyrrole ring precursor.

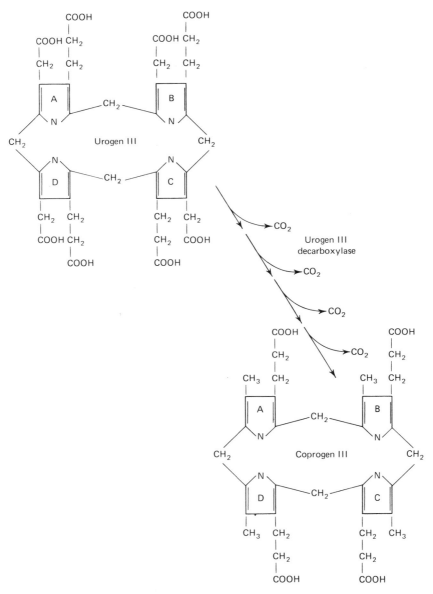

Figure 6.16b   Biogenesis of chlorophyll (*contd.*).

3. Dimerization of two molecules of PBG to give a dipyrrylmethane.[46]
4. Subsequent addition of two more PBG, the last being inserted backward with respect to the first three. The tetrapyrrole cyclizes

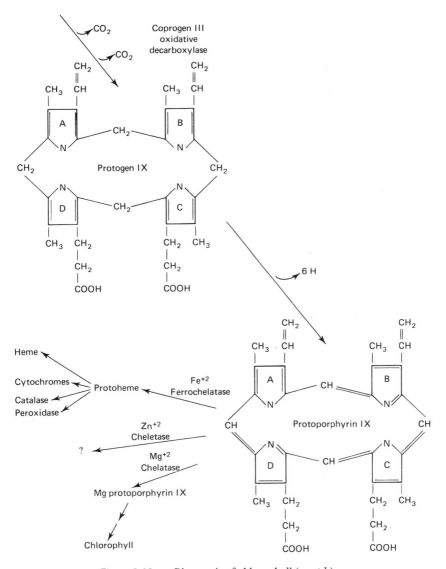

Figure 6.16c   Biogenesis of chlorophyll (*contd.*).

immediately. The reduced cyclized tetrapyrroles are porphyrinogens — easily oxidized to the corresponding porphyrins by air but kept reduced in the biosynthetic pathway.

5. The cyclization product uroporphyrinogen III (= urogen III) is subjected to four decarboxylations of the acetyl side chains.

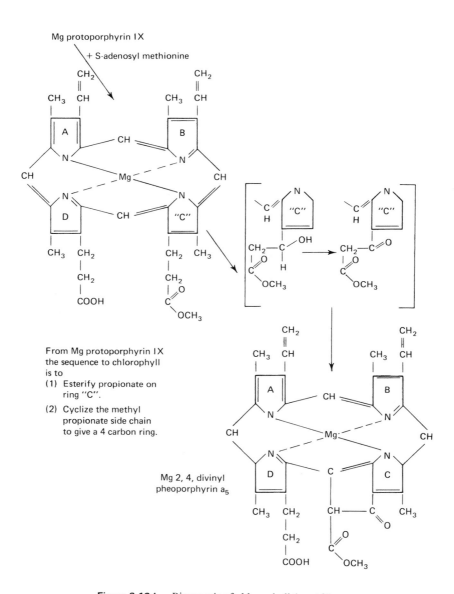

Figure 6.16d   Biogenesis of chlorophyll (*contd.*).

6. The product coprogen III is subjected to oxidative decarboxylation of two propionic side chains to give vinyl side chains.

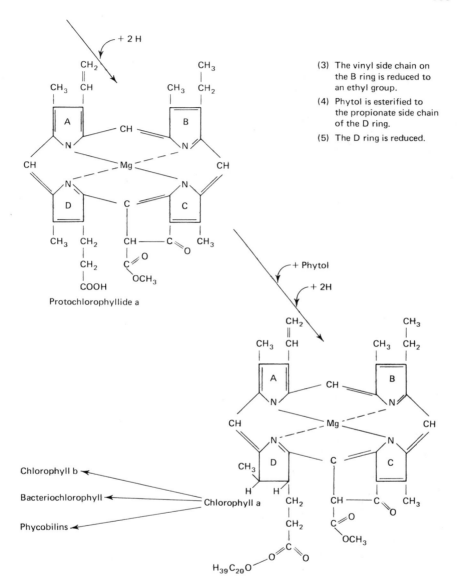

(3) The vinyl side chain on the B ring is reduced to an ethyl group.

(4) Phytol is esterified to the propionate side chain of the D ring.

(5) The D ring is reduced.

Figure 6.16e   Biogenesis of chlorophyll (*contd.*).

7. The product protogen IX is oxidized losing six protons and electrons to give protoporphyrin IX (Fig. 6.16).

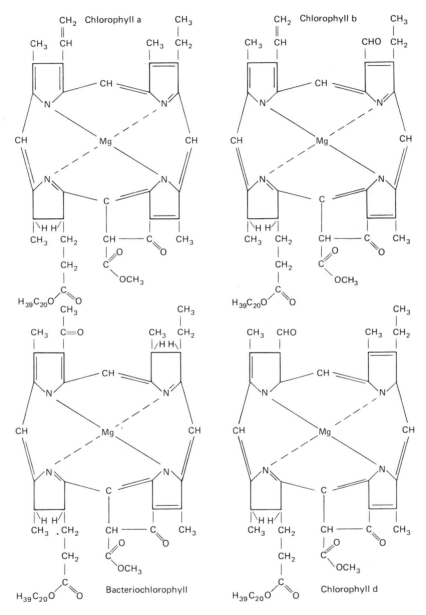

Figure 6.17   Various types of chlorophyll.

Variations are worked in the structure by

   1. Oxidation of methyl group in the B ring to give chlorophyll b.

2. Oxidation of the ethyl side chain of the A ring and reduction of the B ring to give bacteriochlorophyll.

3. Oxidation of the ethyl group to an aldehyde in the A ring to give chlorophyll d (Fig. 6.17).

A major modification of the chlorophyll structure is the opening of the cyclized tetrapyrrole to give a linear chain found in the blue-green and red algae. A proposed mechanism for this is the oxidation of the methine bridge carbon between the A and B rings as shown in Figure 6.18. The production of carbon monoxide from this conversion is firmly established and is rather unique in that CO is not a usual reactant or product in metabolism.[47]

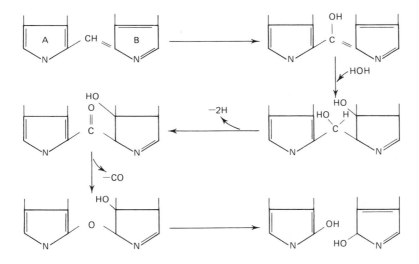

Figure 6.18   Oxidative cleavage of the methine bridge between the A and B rings of chlorophyll to give the linear tetrapyrrole found in phycocyanin and phytochrome.

The sequence of reactions involved in chlorophyll synthesis can be summarized as follows:

1. Form a monopyrrole (PBG from ALA).
2. Stepwise polymerize 4 PBGs to tetrapyrrole.
3. Decarboxylate 4 acetyl side chains.
4. Convert 2 propionyl side chains to vinyl side chains.
5. Remove 6H (porphyrinogens ⟶ porphyrins).
6. Insert Mg^{+2}.
7. Add methyl ester on ring C.
8. Cyclize propionyl side chain on ring C.

9. Reduce vinyl side chain on ring B.
10. Reduce D ring.
11. Add phytol ester on D ring.

The early evidence for this pathway came from Granick's mutant work.[43] Aronoff has found new mutants of *Chlorella* and characterized the accumulated intermediates which suggest that reduction of the vinyl side chain on ring B precedes cyclization and that a hydroxylated intermediate of cyclization may be an intermediate.[48] The work with cell-free systems promises to add further detail here.[45]

The enzymes catalyzing the sequence of reactions for chlorophyll formation are known in a rather fragmentary fashion.

1. The formation of ALA may be accomplished in two ways. The glycine plus succinyl CoA condensation catalyzed by ALA synthetase — the enzyme is known from avian erythrocytes and has been purified from the photosynthetic bacterium *Rhodopseudomonas spheroides.*[49],[50] ALA synthetase has been difficult to detect in higher plant extracts, but a transaminase that aminates alpha ketoglutaraldehyde to give ALA has been found.

Bogorad has shown that actinomycin D, chloramphenicol, and puromycin inhibit chlorophyll synthesis in rapidly greening leaves.[51] ALA partially reverses this inhibition. These observations suggest that chlorophyll synthesis may be controlled by the formation of the enzyme that catalyzes the synthesis of ALA. Chloramphenicol inhibition suggests that a chloroplast ribosome is functioning in the synthesis of this enzyme. In *Chlorella*, the synthesis and breakdown of the enzyme responsible for ALA production appears to be key regulatory mechanism in chlorophyll synthesis.[52]

PBG is formed by the dehydration and condensation of two ALAs catalyzed by the enzyme ALA dehydrase which has been purified from tobacco.[53] This enzyme has also been isolated from *Rhodopseudomonas spheroides* and its mechanism of action studied and some of its control properties have been reported.

2. Polymerization of 4 PBGs to give urogen III is seen in cell-free preparations of *Chlorella* and *Rhodopseudomonas spheroides*. These cell-free reactions produce both urogen I — the isomer in which all 4 PBGs are condensed head to tail — and urogen III — the natural isomer in which the last PBG is put on backward. When the enzyme preparation is heated for 30 minutes at 55° C, only urogen I is formed. This is taken to mean that one enzyme — urogen I synthetase — catalyzes head to tail PBG condensations while the more heat sensitive urogen III cosynthetase either isomerizes the tetramer or directs in the last PBG in reverse. Similar results have been obtained with an enzyme preparation from cultured soy bean

callus.[54] A porphobilinogen deaminase has been purified from wheat germ. In this case, both polymerization and cyclization activities are purified together suggesting a single enzyme while differential responses to inhibitors suggest two distinct catalytic sites.[55]

3. Urogen III decarboxylase has been purified from rabbit erythrocytes and is known from *Chlorella* mutants.

4. The oxidative decarboxylation of coprogen III to give vinyl side chains has been observed as a cell-free reaction, but the enzyme is difficult to purify. Iron deficiency chlorosis is well known and has long indicated a role for iron in chlorophyll synthesis. There is now evidence to suggest that the iron is needed for the conversion of coprogen III to protogen IX.[56,57]

5. The oxidation of protogen IX to protoporphyrin IX goes spontaneously and one must assume that this oxidation is prevented at earlier stages by special attachment to protein or by some other protective mechanism.

6. The mechanism of Mg^{+2} insertion into the tetrapyrrole ring is presumably similar to ferrochelatase action. This enzyme catalyzing the insertion of iron into protoporphyrin IX has been studied in animal tissue and was recently reported in a chloroplast preparation in which it presumably participates in cytochrome synthesis. More puzzling is the discovery of a chelatase in barley that inserts either zinc or copper into porphyrins.[58] No chelatase activity for magnesium ion has been reported yet.

7. The enzyme catalyzing esterification of a methyl group from S-adenosyl methione to the propionyl side chain of the C ring has been studied in preparations of both *Rps. spheroides* and corn leaf.

8., 9. The reactions accomplishing cyclization and the reduction of one of the vinyl side chains are known only from mutant studies and have not been characterized at enzyme level.

10. The reduction of the D ring has been observed as a cell-free reaction. This reduction is a light-dependent reaction in some algae and most higher plants (conifers are able to turn green in the dark). The conversion of protochlorophyllide to chlorophyllide occurs on a protein – protochlorophyllide holochrome. This conversion may be observed spectroscopically when etiolated leaves or a homogenate of etiolated leaves is illuminated. Precise spectral measurements reveal intermediate stages of conversion indicating that protochlorophyllide is reduced; then its binding to the protein alters or the conformation shifts in a presently unknown way to alter the absorption spectrum (Fig. 6.19).

Kinetic studies of the greening process allow some interesting inferences:

    a. The rate is directly proportional to the light intensity indicating that only one photochemically excited molecule is involved.

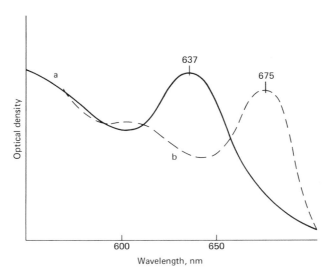

Figure 6.19   Absorption spectra of solutions of (a) protochlorophyllide holochrome and (b) chlorophyllide holochrome from etiolated red
kidney bean. The latter was obtained by illuminating the former
solution for 30 seconds. [From Capon, B., and L. Bogorad,
Botan. Gaz., **123**, 285 (1962).]

    b. Although the reaction occurs below $0°$ C, the rate varies at low
temperatures indicating that the reaction is not entirely photochemical.

    c. Since the rate is independent of viscosity, the reaction is mono-
molecular.

Since the new hydrogens in the D ring of chlorophyllide cannot be
labelled with deuterium when the protochlorophyllide is photoconverted
in D_2O, one might imagine that light absorption by the protochlorophyllide causes a restricted collision within the holochrome complex where the
protein serves as a hydrogen donor for reduction of the chromophore.

There are several reports of the purification of protochlorophyllide
holochrome to homogeneity from etiolated leaves. The molecular weight
of the protein is 550,000 to 600,000 and the other physical properties
reported for these preparations are in good agreement. Treatment of the
protochlorophyllide holochrome from barley yields a subunit of molecular
weight 63,000 and this is reported to undergo photoconversion.[59]

There is increasing spectroscopic evidence for multiple steps in the
photoconversion and subsequent dark rearrangement of the chromophore
on the protein.[60] Two distinct forms of chlorophyll-protein complexes
may form during greening of etiolated tissue.[61]

11. The addition of the phytol side chain is probably catalyzed by the enzyme chlorophyllase which will split as well as form this ester.[62] The enzyme is at its highest activity in young plants synthesizing lots of chlorophyll. A *Chlorella* mutant that excretes chlorophyllide a (chlorophyll without the phytol) has been found.[63] This mutant has only 20% of the "wild type" level of chlorophyllase and has no carotenoids. This suggests a defect in isoprenoid synthesis with chlorophyllase as a substrate induced enzyme. This mutant situation would imply some more intricate controls of chlorophyll synthesis than simply regulation of ALA synthetase. It is also noteworthy that this mutant contained no chlorophyllide b suggesting that chlorophyll a is converted to chlorophyll b (an oxidation of the methyl side chain on ring B to an aldehyde) only after phytol esterification. Several cell-free systems from higher plants for the conversion of chlorophyll a to chlorophyll b have been reported. One of these preparations requires NADP supplementation[64] while the other suggests that a photoinduction may be required to activate chlorophyll b synthesis.[65]

Other steps for modification of chlorophyll are not known at the enzyme level. Troxler has developed isotope evidence for the formation of the linear tetrapyrroles in the phycobiliproteins of the alga *Cyanidium*. On feeding C^{14} ALA, copro III and phycocyanin – the linear tetrapyrrole which in these algae replaces chlorophyll b – contained equal specific activity which was fourfold above PBG and twice the activity of ALA. These data indicate the sequence

$$ALA \longrightarrow PBG \longrightarrow \longrightarrow coprogen\ III \longrightarrow \longrightarrow \longrightarrow phycocyanin$$

indicating that the alga first forms the cyclic compound which is then opened to the linear tetrapyrrole. There is evidence for a parallel synthesis of the protein suggesting a coordinated regulation of both polypeptide and pigment synthesis in the formation of phycobiliproteins.[66]

GENERAL REFERENCES

Hitchcock, C., and B. U. Nichols. *Plant Lipid Biochemistry*. New York: Academic Press, 1971.

Kolattukudy, P. E. "Biosynthesis of Cuticular Lipids," *Ann. Rev. Plant Physiol.*, **21**, 163 (1970).

Lascelles, J. *Tetrapyrrole Biosynthesis*. New York: Benjamin Publishing Co., 1964.

Nichols, B. W., and A. T. James. "Acyl Lipids and Fatty Acids in Photosynthetic Tissues," *Progress in Phytochemistry*, **1**, 1 (1968).

Porter, J. W., and D. G. Anderson. "Biosynthesis of Carotenes," *Ann. Rev. Plant Physiol.*, **18**, 197 (1967).

7 PHOTOSYNTHESIS: PHYSICS, PHOTOSYSTEM I

He had been eight years upon a project of extracting sunbeams out of cucumbers, which were to be put into vials hermetically sealed, and let out to warm the air in raw, inclement summers. He told me, he did not doubt in eight years more that he should be able to supply the Governor's gardens with sunshine at a reasonable rate.

JONATHAN SWIFT, *Gulliver's Travels*, Book III, "A Voyage to Laputa," 1726.

Photosynthesis supplies the reduced carbon compounds on which all heterotrophic life is dependent. As noted in Chapter Four, photosynthetic carbon metabolism involves some unique enzymatic reactions, but these are not different in kind from reactions that occur in heterotrophic cells. The distinctive feature of photosynthetic organisms is their ability to convert radiant energy into chemical potential. The energy conversion process has attracted much effort to understanding photosynthesis in terms of physics. At present, the biochemical details of the energy conversion process are becoming progressively clearer. The rate of clarification has been slow since the light-dependent chemical activities are tightly bound to chloroplast membrane structures. Conventional biochemistry is much better equipped to understand events that take place in solution. As methods develop for understanding solid-state processes in biological materials, photosynthesis should become much less mysterious.

7.1   QUANTUM EFFICIENCY

A description of photosynthesis might well begin with a fundamental physical parameter — the quantum requirement or its reciprocal, the quantum yield of this process. If one knew the minimum number of quanta actually required for photosynthesis, this number would place strict limitations on interpretation of the chemical mechanism. With respect to the overall process of photosynthesis, the number of quanta actually required to fix a molecule of CO_2 or the efficiency of the plant in utilizing light energy is a matter amenable to experimental measurement.

The measurement of the quantum requirement is of general utility in studying all photochemical processes. Light can do almost anything, i.e., all photochemical reactions are possible if you accept a low photochemical efficiency. Put another way, all rugs will fade if left in sunlight long enough. Since everything is possible with photochemistry, not everything is important. In a biological system, the importance of the light response must be judged in terms of its quantum requirement — this is especially true in relating an *in vitro* photochemical reaction to the photo response of the intact organism.

The theoretical case for calculation of the quantum requirement of photosynthesis goes as follows. The wavelength of 678 nm light is chosen since it is absorbed at the red peak in the chlorophyll absorption spectrum and since it will drive photosynthesis in intact plants.

$$\lambda = 678 \text{ nm} = 6.78 \times 10^{-5} \text{ cm}$$

The energy of a quantum of this light can be obtained from the equation

$$E \text{ (energy of a quantum)} = \frac{h \text{ (Planck's constant)} \times c \text{ (velocity of light)}}{\lambda \text{ (wavelength)}}$$

which in real numbers is

$$E = \frac{6.62 \times 10^{-27} \text{erg sec} \times 2.99 \times 10^{10} \text{cm/sec}}{6.78 \times 10^{-5} \text{cm}}$$

$$= \frac{19.79 \times 10^{-17}}{6.78 \times 10^{-5}} = 2.9 \times 10^{-12} \text{ergs/quantum}$$

You might reasonably expect one quantum to directly affect one molecule, which is Einstein's law of photochemical equivalence. Talking about one quantum or one molecule in the current world is an analytical fantasy and one usually works with an Avogadro's number of molecules or quanta. An Avogadro's number of quanta (or a mole of quanta) is called an Einstein and the energy value obtained above is converted to this unit.

$$2.9 \times 10^{-12} \text{ ergs/quantum} \times 6 \times 10^{23} \text{ quanta/mole} =$$
$$17.4 \times 10^{11} \text{ ergs/Einstein}$$

To further approach the chemical world, convert ergs to calories

$$\frac{17.4 \times 10^{11} \text{ ergs/Einstein}}{4.185 \times 10^{7} \text{ergs/calorie}} = 4.1 \times 10^{4} \text{ calories/Einstein}$$

or 41,000 calories are available in a mole of quanta at $\lambda = 678$ nm.

Thermochemical data give the free energy required for formation of glucose from CO_2 and water as 708,000 calories with approximate corrections for *in vivo* conditions. Taking $\frac{1}{6}$ of this value gives 118,000 calories required to fix one mole of CO_2. Thus a theoretical minimum of $\frac{118,000}{41,000} = 3$ quanta are required to fix one CO_2.

How does one measure the real quantum requirement in an experimental context? Pure light of a specified wavelength is obtained with interference filters, a prism, or a grating monochrometer. The available energy at the wavelength in question is measured with a thermopile or a photoelectric device calibrated against a standard lamp. The National Bureau of Standards will supply a standard lamp that can be depended on to supply a specified number of quanta under controlled conditions. With a calibrated detector, one gets a signal translatable into ergs/unit of absorbing area = cm^2/unit time = seconds. The reaction vessel is then placed in the light beam and unless the sample is totally absorbing all of the light, the light transmitted through the sample is subtracted from the incident energy to give a rough measure of the light absorbed. To improve this value, one must correct for light reflected off the vessel surface, for light scattered by the biological sample which is rarely in true solution, and for light absorbed by non-reactive pigments. The scattering and reflection problems are best solved by using an integrating sphere or bolometer. This device simply measures the non-absorbed light in all directions around the sample. As to light absorbed by non-reactive pigments, one simply tries to have the best biological material possible. Finally, one gets an estimate of the number of quanta absorbed by the sample for a period of time during which some chemical change is also measured. The number of quanta required to change one molecule is the *quantum requirement* and the number of molecules changed per quantum absorbed (usually less than one) is the *quantum yield*.

In estimating the quantum requirement for photosynthesis, algae are usually used since they are easily handled. One measures either O_2 production or CO_2 consumption resulting from the absorption of a known number of quanta. A critical problem is what to do about respiratory consumption of O_2 and production of CO_2 — do these processes continue unaffected during photosynthesis? Again, various corrections for respiratory activity can be devised. The measured values for the quantum requirement have been a point of controversy. Several laboratories have reported values of 8 or more quanta required per CO_2 fixed, indicating an efficiency of $(\frac{118,000}{8 \times 41,000} = 0.36)$ 36%. Warburg found that 3 or fewer quanta were sufficient, suggesting 100% efficiency with a partial energy contribution from a respiratory back reaction. This divergence of measurements is an enigma. Warburg claimed to have superior biological material, and his measurements had been witnessed by reputable observers. However, independent measurements consistently give the higher quantum requirement.

For the cell-free partial reactions of photosynthesis, chloroplast NADP reduction, or synthesis of ATP, the quantum requirements tend to be high. Usually when such reactions are first discovered, the requirement is absurdly high but improved reaction conditions generally bring the rate up and the quantum requirement down.[1] The current values for the quantum requirements for these cell-free reactions are in the right order of magnitude. Tearing things out of the cell could cause many kinds of changes that would lower photosynthetic efficiency.

7.2   ACTION SPECTRA, ENHANCEMENT, PHOTOSYNTHETIC UNITS

A simpler physical measurement is that of action spectra. One measures the amount of action — CO_2 fixation, O_2 production, NADP reduction, etc. — as a function of wavelength to discern which pigments can collect radiant energy for use in chemical work. In plants, most of the chloroplast pigments — chlorophyll a and b or phycobiliprotein, if present, and carotenoids are functional in light harvesting. Flavonoids and anthocyanins, localized in the vacuole or cytoplasm of some leaf cells, are non-functional in photosynthesis. Figure 7.1 shows the correspondence between the absorption spectrum and the action spectrum of a red alga. Exceptions to the rule of chloroplast pigments functioning in photosynthesis are known. In both bacteria and algae, the carotenoids may vary in their effectiveness in contributing light energy to photosynthesis. Phycobilins and even chlorophyll a need not be completely effective in situations in which the cell overproduces these pigments or imperfectly incorporates them into the photosynthetic structure.

A very influential observation was made in measuring the quantum yield as a function of wavelength in a number of plant species. The quantum yield is reasonably constant for all wavelengths until one approaches 700 nm and then there is a sharp drop in quantum yield as shown in Figure 7.2.

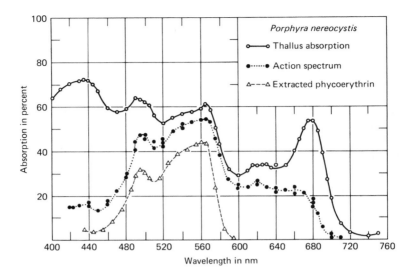

Figure 7.1   The absorption and action spectra of the red alga, *Porphyra nereocystis*. Note that the action spectrum is the sum of the absorption spectra of phycoerythrin and of chlorophyll a. [From Haxo and Blinks, *J. Gen. Physiol.*, **33**, 389 (1950).]

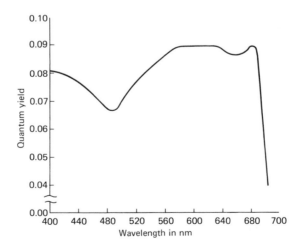

Figure 7.2   The quantum yield of *Chlorella* photosynthesis as a function of the wavelength of light. Note the precipitous drop in yield when light of wavelengths which are absorbed only by chlorophyll a is used. [From Emerson, R. and Lewis, C. M., *Amer. J. of Botany*, **30**, 165 (1943).]

The point to note here is that light is absorbed by the plant — by chlorophyll a — at wavelengths above 700 nm, but this light is not used as efficiently as light of shorter wavelengths. The other point to note is that a quantum of blue light (e.g., 420 nm), which has nearly twice the energy of a quantum of red, gives no greater yield of photosynthesis. A blue quantum at 420 nm will apparently raise a loosely held π electron from the resonance cloud in a highly conjugated chlorophyll to the second excited state. This electron must drop to the first excited singlet level with dissipation of energy as heat before it is ready to do chemistry. From the first excited singlet, which also can be reached by absorption of a red quantum at 680 nm, the electron can return to the ground state with loss of energy, either by useful chemical work of photosynthesis or by fluorescence or heat dissipation such as occurs with isolated chlorophyll in solution.

Returning to the phenomenon of the red drop of quantum yield above 700 nm, Emerson, who discovered it, noted that this was the only region in the entire plant absorption spectrum where chlorophyll a was the only absorbing pigment.[2] In all other regions some other pigment in addition to chlorophyll a was absorbing light. That these accessory pigments could transfer the absorbed light energy to chlorophyll a was well known. Resonance transfer of energy between molecules demands an overlap in absorption spectra so that fluorescent emission of one pigment can be collected by the next pigment and a minimal physical proximity of the two pigments, which is guaranteed by gluing the pigments close together in the chloroplast. Thus the accessory pigments could feed energy into chlorophyll a, but when the chlorophyll a alone absorbed quanta, the energy was poorly used in photosynthesis.

Emerson's discovery of the "red drop" led him to a discovery of "enhancement." Having noted the inefficiency of light absorbed only by chlorophyll a, Emerson tested the effect of adding to the beam of red light a weak beam of light absorbed by the accessory pigment and found that the combination of beams gave a rate of photosynthesis which was more than the sum of the rates when the beams were used separately. Thus the rate of photosynthesis is greater when both chlorophyll a and the accessory pigment are illuminated simultaneously.

Photosynthesis at 710 nm		10
" " 650 "		43.5
" " 650 + 710 nm		72.2

It may be useful to calculate enhancement as a ratio of the rate in both wavelengths simultaneously presented over the sum of the rates in each wavelength:

$$\frac{72.2}{10 + 43.5} = 1.35$$

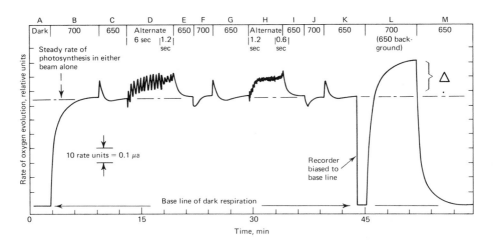

Figure 7.3   A measurement of enhancement of photosynthetic oxygen
production by *Chlorella*. The 700 nm and the 650 nm light
beams were adjusted to yield equal steady state rates. A surge
of enhanced activity occurs on switching from 700 to 650 nm
light (B → C) and this can be sustained by rapidly alternating
the beams (D and H) indicating the persistence of a product of
one light reaction which enhances the other light reaction. Enhancement by simultaneous exposure to both beams is shown
(L vs M). [From Meyers, J. and French, S., *Plant Physiol.*, **35**,
963 (1960).]

When one varies the wavelength of the supplementary beam, the increase in
photosynthesis gives an action spectrum corresponding to the absorption
spectrum of the accessory pigment. It became apparent that both beams need
not be presented simultaneously but could be separated by a short interval of
time. Blinks measured steady-state O_2 production in alternately flashing beams
and found that intervals of 0.6 to 1.6 seconds separation between flashes still
allowed enhancement.[3] An elegant demonstration of these observations is shown
in Figure 7.3. Clearly there is dark chemistry involved in the summation of the
two photochemical events. The Blinks chromatic transient experiment is easily
illustrated as follows. Photosynthetic oxygen production was measured with a
fast response platinum electrode. Two beams of light were used sequentially and
the intensity of each beam was adjusted to give the same rate of O_2 evolution
(n.b. the light intensities here as in the Emerson enhancement experiments are
below light saturation). When the irradiating beam was quickly switched from
one wavelength to another, Blinks noted a transient increase in O_2 evolution. In
700 nm light the algae remembered (for a fraction of a second) the previous
exposure to light of a different wavelength. When the 700 nm beam absorbed by
chlorophyll a only was held constant and the wavelength of the other beam was

varied, Blinks found that the oxygen spike or chromatic transient had an action spectrum identical to the absorption spectrum of the accessory pigment (chlorophyll b in green algae, carotenoids in brown algae, phycobilins in red or blue-green algae). Again, the evidence suggested two distinct photochemical acts separated from one another by dark chemical reactions. Consistent with this was the old observation that all plants that evolved oxygen had two major photosynthetic pigments — chlorophyll a and either chlorophyll b or some substitute like phycobilin or a carotenoid. Note that not all carotenoids can be considered major accessory pigments. Green leaves contain both carotenoids and chlorophyll b, and although the carotenoids may harvest some light energy for photosynthesis, chlorophyll b is the major accessory pigment.

The early simplified view was to think that for oxygen evolution, one photo act was accomplished exclusively by chlorophyll a and the other by chlorophyll b or the accessory pigment. The error of this was soon apparent when cell-free preparations of blue-green algae, which had been thoroughly depleted of the water soluble accessory pigment phycocyanin, were found to be competent in O_2 evolution. Mutants of green algae and of higher plants blocked in the synthesis of chlorophyll b were found to be competent in complete photosynthesis. The way out of this dilemma was to assume that chlorophyll a was active in both photosystems — most of it in Photosystem I, but a bit in Photosystem II.

Experimental support for the participation of chlorophyll a in both photosystems came from the work of Duysens.[4] He examined the fluorescent light emission of chlorophyll a in intact algae. When a pigment absorbs light energy and cannot use it for chemical work, some of the energy is lost by emission of light quanta of a longer wavelength. If a plant is prevented from doing photosynthesis by lack of CO_2 or by inhibitors blocking enzymatic utilization of the absorbed light energy, some of the energy will be lost by chlorophyll as fluorescence. Duysens chose a red alga in which chlorophyll a (680 nm) and phycoerythrin (550 nm), the major accessory pigment of Photosystem II, have absorption peaks that are well separated. He found that irradiation of the chlorophyll a band in the absence of CO_2 resulted in little chlorophyll a fluorescence, but that irradiation of the phycoerythrin gave a pronounced chlorophyll a fluorescence. This led Duysens to postulate that there were two forms of chlorophyll a — a large pool of non-fluorescent chlorophyll a associated with Photosystem I and a smaller pool of fluorescent chlorophyll a in Photosystem II that can accept energy (by physical resonance transfer) from phycoerythrin and then emit the energy as fluorescence. The action spectrum of chlorophyll a fluorescence in intact plants was identical to the absorption spectrum of the accessory pigment.

Thus action spectra measurements have indicated the existence of two photosystems and have given a rough idea of the pigment composition of these two systems.

Although enhancement is generally observed for the photosynthesis of intact cells, there is controversy concerning its occurrence in cell-free reactions. Although there are several reports of enhancement of NADP reduction by isolated chloroplasts, Arnon has failed to observe this under circumstances in which he can measure enhancement of CO_2 fixation.[5] These experiments have led him to postulate that chlorophyll a is used mainly for ATP generation and that chlorophyll b is responsible for NADP reduction. The enhancement of NADP photoreduction by cell-free systems may depend on conditions of measurement that need to be better controlled and better understood.[6,7] Warburg's high-efficiency photosynthesis in intact cells precludes the need for the kind of cooperation of photosystems indicated by the enhancement studies.

There is one more physical approach to the problem of photosynthesis which is germane to a biochemical understanding of the process. At the turn of the century, Blackman had established that the yield of photosynthesis in saturating light was a function of temperature, a result that could only mean that there were non-photochemical dark reactions of photosynthesis. Emerson and Arnold devised a physical experiment to learn the relative size of the photochemical and dark processes before there was any clue to the nature of the dark reactions. They used a bright flash of white light and varied the dark time between flashes to get the maximum efficiency of photosynthesis from a sample of algae. The notion was to allow dark time for the slow chemical reactions to drain off the products of the saturating flash of light. They found that a repetitive 10^{-5} second flash would give the maximum yield of oxygen if the flashes were interspersed with 0.04 seconds of darkness. Under these conditions of maximum efficiency, one molecule of oxygen is evolved per 2,500 chlorophyll molecules present, suggesting that the chlorophyll is present in great excess over the dark reaction machinery.[8] A computational approach to this problem is made by calculating from the number of chlorophylls present and from the number of quanta absorbed in very dim light that a given chlorophyll molecule would absorb a quantum on the average of once an hour, yet the steady-state rate of photosynthesis is established very quickly after turning the light on, regardless of the light intensity. These considerations led to the notion of a photosynthetic unit in which a large number of chlorophyll molecules were painted out on an absorbing surface in order to catch any quanta that came along. Once energy was captured on the absorbing surface, it could be transferred by known physical processes to a trap or chemical reaction center to initiate dark reactions. The notion is appealing since it saves the plant from the necessity of making a lot of enzymes for every chlorophyll when these enzymes would have to sit around doing nothing most of the time while waiting for a rare quantum absorption.

7.3   THE REDUCTION OF CO_2 VIA NADP

Photosynthesis is obviously an oxidation-reduction process — carbon dioxide is reduced to carbohydrate and a biochemical description of this process centers on the individual chemical identities of the participants. In order to examine the individual redox components of the photosynthetic electron transport chain, the currently fashionable but not quite universally accepted Z scheme, illustrated in Figure 7.4, will be used. This scheme is still a working hypothesis with considerable basis in fact, but it is probably not as secure as the mitochondrial respiratory chain.

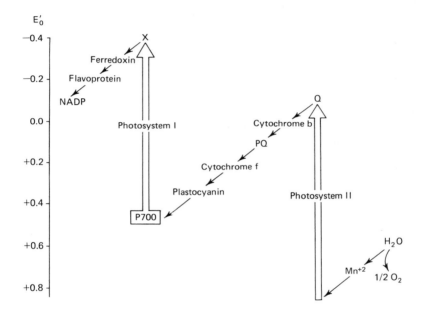

Figure 7.4   A diagrammatic representation of the photosynthetic electron transport sequence. The two photosystems drive electrons against redox potential gradient while the dark reactions around them are exergonic and spontaneous.

It has been evident for some time that the business of illuminated green plants was to reduce carbon dioxide to carbohydrate with concomitant production of oxygen. R. Hill opened up the biochemistry of this process by demonstrating the light-dependent oxygen production by a cell-free preparation of higher plant leaves. Chloroplasts were easily recognized as the unique photosynthetic organelles in the plant cell since microscopic examination of plant cells shows that all the chlorophyll is in the chloroplasts. Hill isolated chloroplasts and when these particles were illuminated in the presence of a suitable electron acceptor, the electron acceptor was reduced and oxygen was evolved with a stoichiometry suggesting that water was the source of both the reducing power and the oxygen. The first equations describing this — the Hill reaction — were written

$$H_2O + A \xrightarrow[\text{chloroplasts}]{\text{light}} AH_2 + \tfrac{1}{2}O_2$$

where A was a "suitable" oxidant. Ideally, the suitable oxidant would be carbon dioxide since this is what the intact plant reduces, but the rather complex and weakly held enzyme system for carbon dioxide reduction had been lost from the chloroplast during isolation. Failing carbon dioxide reduction, it would be "suitable" if the oxidant A should be something which, when reduced, would be a strong enough reducing agent (have a sufficiently negative reducing potential) to reduce carbon dioxide. Unfortunately, the experimentally "suitable" oxidants, i.e., things that worked, were of rather too positive a redox potential and, when reduced by chloroplasts, would not reduce carbon dioxide. Ferricyanide, benzoquinone, and indophenol dyes were able to serve as Hill oxidants.

By analogy to other carboxylation reactions followed by reduction of the carbon dioxide fixed in bacterial and animal systems, it was guessed that the chloroplast should reduce pyridine nucleotide in a Hill reaction. With the elucidation of the Calvin cycle using $C^{14}O_2$ fixation in intact illuminated plants, the sequence for carbon reduction was recognized as that shown in Figure 7.5. NADPH, rather than NADH, was strongly suggested by the prevalence of NADPH in biosynthetic reactions. This suggestion of NADPH as the coenzyme was supported by the appearance of an NADP-specific glyceraldehyde phosphate dehydrogenase activity in etiolated plant tissue on transfer from darkness to light. Thus the NADP-linked enzyme was associated with the appearance of photosynthetic competence. The only exception seems to be in most photosynthetic bacteria in which the glyceraldehyde phosphate dehydrogenase shows a persistent, absolute specificity for NAD.

7.4   FLAVOPROTEIN AND FERREDOXIN

As far as the isolated chloroplasts were concerned, these organelles did not reduce NADP in the light and they did not even contain a significant amount of NADP or much else that was easily water soluble. This simply meant that the chloroplasts were leaky and lost their loosely held, water soluble constituents during isolation. In 1956, San Pietro and then Arnon demonstrated the net reduction of pyridine nucleotides by illuminated chloroplasts when supplemented with soluble protein from leaf material. Purification of the soluble protein required for chloroplast NADP photoreduction led to the isolation of a low molecular weight, iron protein — ferredoxin. This protein is similar to the bacterial ferredoxin first discovered by Mortenson as a participant in Clostridium nitrogen fixation. Ferredoxins were then crystallized from a number of plant

121

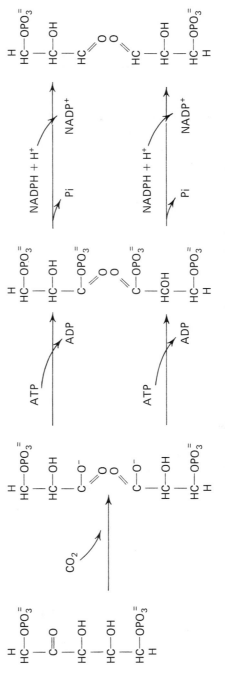

Figure 7.5 The immediate requirements for reduced pyridine nucleotide and ATP in photosynthetic carbon dioxide fixation.

and bacterial sources and some of their properties are compared in Table 7.1. Ferredoxins from many bacteria, several algae, and a number of higher plant species have been or are being sequenced and a comparison of primary structures already shows interesting implications concerning the evolution of this molecule.[9] Ferredoxins have characteristic absorption spectra and Figure 7.6 shows the absorption spectrum of a typical photosynthetic ferredoxin. The ESR signal of reduced ferredoxin is seen only at low temperature and is not frozen out at $2°K$ which is taken to mean a strong interaction between irons through one or more ligands, likely the sulfur.[10,11] This iron-sulfur moiety represents a new type of protein constituent that characteristically releases H_2S at acid pH.

TABLE 7.1   A COMPARISON OF FERREDOXINS FROM DIFFERENT ORGANISMS

	Clostridium	*Chlorobium*	*Chromatium*	*Spinach*
Number of amino acids	57	55	81	97
Fe/mole protein	7	5	6	2
H_2S/mole protein	7	5	6	2
E_0' at pH 7.5	−0.47		−0.49	−0.43
Low temp. ESR signal, g	1.89–1.96	−	−	1.98–1.96

The plant and bacterial ferredoxins are further distinguished by the fact that two irons are reduced and oxidized in the bacterial proteins while the plant ferredoxins function as single electron carriers. That is, two moles of reduced spinach ferredoxin are required to reduce one mole of NADP. Another functional distinction is apparent in the very poor interchangeability of bacterial and plant ferredoxins. While the plant ferredoxin works well in the bacterial hydrogenase assay, bacterial ferredoxin is poorly reduced by chloroplasts.

At this point it is clear that ferredoxin isolated from green plant leaves and presumably loosely held in the chloroplast is required for NADP reduction. Many lines of evidence point to participation of a second enzyme that transfers electrons from reduced ferredoxin to NADP. Arnon isolated spinach ferredoxin and used a bacterial hydrogenase preparation to reduce the ferredoxin with hydrogen gas. NADP reduction occurred only when chloroplasts were added. The system − H_2 + hydrogenase + spinach ferredoxin − gave reduced ferredoxin. Chloroplasts, in the dark, supplied a catalyst to transfer electrons from reduced ferredoxin to NADP. This catalyst was purified and identified as a flavoprotein which had previously been isolated on the basis of its ability to transfer electrons from NADPH to indophenol dye. This enzyme had also been isolated on the basis of its activity in catalyzing the transfer of electrons from

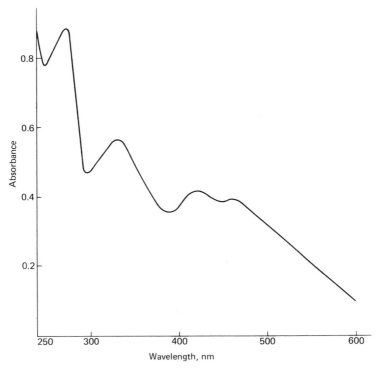

Figure 7.6   An absorption spectrum of a "higher plant" type ferredoxin.

NADPH to NAD. The flavoprotein bears the official name of ferredoxin-NADP oxidoreductase. Since the enzyme activity remains in spinach chloroplasts after many washings with isotonic sucrose, its role between ferredoxin and NADP remained obscure as long as illuminated chloroplasts were used to generate reduced ferredoxin. When purified and crystallized, the flavoprotein from spinach was found to contain one FAD, no metal, and an essential sulfhydryl group, and it had a molecular weight around 40,000.[12] The ferredoxin-NADP oxidoreductase will reduce NAD, but the K_M for NAD is two orders of magnitude higher than for NADP. The flavoprotein reduces NADP stereospecifically on the A side and oxidizes it on the same side in the diaphorase reaction. This flavoprotein has been isolated from a blue-green alga in which it is very loosely held to the photosynthetic structure and the protein from this source appears to be similar both physically and functionally to the higher plant enzyme. In photosynthetic bacteria, NAD reduction by illuminated chromatophore preparations has been demonstrated.[13] As noted earlier, the triose phosphate dehydrogenase in photosynthetic bacteria is NAD specific and Ogren has demonstrated that NAD, rather than NADP, is reduced in *R. rubrum* when shifted from darkness to light.[14] An apparent exception occurs in *Rhodo-*

pseudomonas palustris in which Kamen has found an NADP-linked flavo-protein.[15] In the other photosynthetic bacteria the NADH generated in photosynthesis may be used to reduce NADP in a conventional transhydrogenase reaction

$$NADH + NADP \rightleftharpoons NAD + NADPH$$

as described for *Chromatium* or perhaps through an energy-linked transhydro-enase seen in *R. rubrum*. Perhaps this latter activity might better be considered as a model for understanding energy transfer than as a physiological shuttle for reducing power. The scheme depicted in Figure 7.7 summarizes the experiments on the energy-linked transhydrogenase in *R. rubrum*.[16] The stoichiometry is one ATP per NADPH and the system is analogous to the energy requiring reverse electron transport in mitochondria.

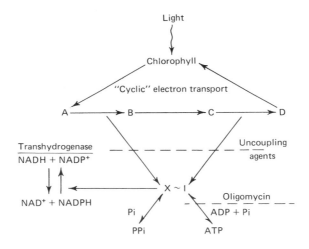

Figure 7.7   The energy-linked transhydrogenase activity of *Rhodospirillum rubrum* chromatophores. Light-induced electron transport drives the synthesis of a high-energy intermediate which can supply energy for ATP or pyrophosphate synthesis or for trans-hydrogenation. [From Keister, D. L. and Yike, N. J., *Biochem. Biophys. Res. Commun.*, **24**, 519 (1966).]

Recently, several observations have turned up on the interaction of ferredoxin with the ferredoxin-NADP oxidoreductase.[17] Neumann, in measuring the reaction

$$NADPH + \text{oxidized indophenol dye} \longrightarrow NADP + \text{reduced dye}$$

catalyzed by ferredoxin-NADP oxidoreductase, found that ferredoxin inhibited this activity.[18] On mixing ferredoxin and the flavoprotein, there is a spectral shift that indicates interaction between the proteins. Shin and San Pietro have reported evidence from optical rotatory dispersion measurements for this protein-protein interaction.[19] This may prove an interesting system to study in solution the interaction of functionally adjacent catalysts.[20],[21],[22]

There are several possibilities of physiological regulation in this area of the electron transport scheme. Fredricks has found that some unidentified compounds in boiled spinach extract inhibit ferredoxin-NADP oxidoreductase but not cyclic phosphorylation.[23] This may be a physiological regulator of electron flow, but a specific identification and measurements of the rise and fall in regulator commensurate with the rise and fall of the regulated activities are needed. In blue-green algae, Smillie found an FMN-containing protein that would substitute for ferredoxin in transferring electrons from the illuminated chloroplast to the FAD-containing ferredoxin-NADP oxidoreductase.[24] At about the same time, the DuPont group found that iron-deficient nitrogen-fixing bacteria would produce an FMN-containing protein that replaced bacterial ferredoxin in the nitrogen-fixing system and which they called flavodoxin. Smillie named the algal protein phytoflavin and Trebst and Boethe subsequently showed that the appearance of phytoflavin in blue-green algae is regulated by the availability of iron.[25] Phytoflavin has been isolated from *Chlorella*.[26]

As a chloroplast component, NADP itself was in doubt for some time. This coenzyme simply leaks out of chloroplasts during isolation. Ogren established the NADP content of higher plant chloroplasts by non-aqueous isolation techniques and has confirmed this with the newer chloroplast isolation methods which preserve the outer chloroplast membrane.[14] Ogren found that the NADP in the *in vivo* chloroplast is converted to NAD when the plant is taken out of the light. There is abundant phosphatase activity *in vitro* to account for this dephosphorylation. When the plant is returned to light, the chloroplast NAD is promptly phosphorylated to NADP. Although a conventional NADP kinase is present in the cytoplasm, this activity is absent from the chloroplast and an *in vitro* system for NAD $\longrightarrow$ NADP conversion has not been found.

Although ATP and ADP can probably pass freely in and out of the chloroplast,[27] there is some dispute about the penetration of NADP. Heber and Santarius have presented strong evidence that NADP, NADPH, and phosphate cross the chloroplast membrane *in vivo* with great difficulty.[28] So, can the chloroplast export reducing power to the cytoplasm in addition to meeting the demands of carbon dioxide fixation within the chloroplast? It is known that there are glutamic and aspartic dehydrogenases both inside and outside the chloroplast and that glutamate, aspartate, alpha ketoglutarate, and oxaloacetate are all freely permeable to the chloroplast membrane.[29] By reducing the alpha keto acids inside the chloroplast, allowing them to diffuse out to the cytoplasm,

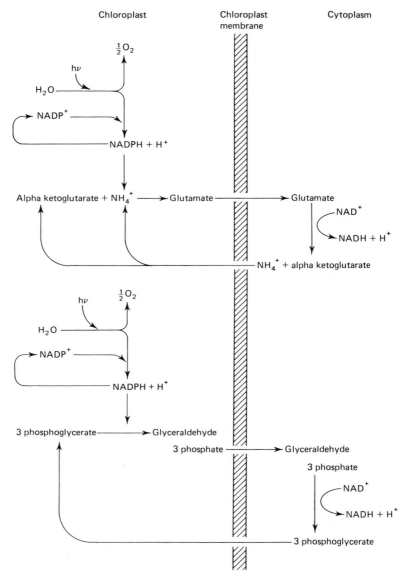

Figure 7.8   Shuttle mechanisms for the export of photosynthetic reducing power across the chloroplast membrane.

and then oxidizing them, the export of reduced pyridine nucleotide is accomplished. The alpha keto acids might diffuse back into the chloroplasts and be available for the next cycle of export. In addition, there appears to be a glyceraldehyde-3-phosphate — phosphoglyceric acid shuttle with appropriate enzymes on each side of the chloroplast membrane as visualized in Figure 7.8.

That NADP may not be the only physiological oxidant of chloroplast reducing power is suggested by the cell-free experiments of Losada.[30] He has found a nitrite reductase activity directly linked to reduced ferredoxin in higher plants and similar reactions have been found in blue-green algae.[31] Similarly, the spinach chloroplast sulfite reductase, a low-potential heme protein, is reduced by reduced ferredoxin.[32] Finally, Arnon has indicated that reduced ferredoxin is the specific reductant in his reverse Krebs cycle for carbon dioxide fixation in photosynthetic bacteria.[33] Here, reduced ferredoxin is seen to participate in the carboxylation of acetyl CoA and and succinyl CoA to pyruvate and alpha ketoglutarate, respectively. Figure 7.9 summarizes the reactions catalyzed by this part of the chain.

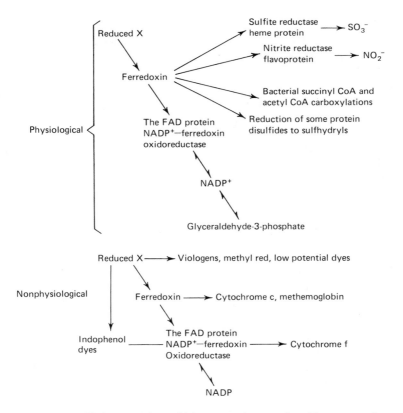

Figure 7.9   Various reactions which remove electrons from Photosystem I.

7.5   A PRIMARY REDUCTANT OF PHOTOSYSTEM I

The current problem with the ferredoxin reducing substance X is its chemical identity. In the Z scheme, X is used to denote that substance which accepts electrons from the photo act and passes them to ferredoxin. Two different spectroscopic techniques have yielded information about temperature-insensitive, light-induced changes in chloroplasts that are associated with Photosystem I. Hiyama and Ke found light-induced absorbance changes at 430 nm in chloroplast preparations.[34] These changes show an apparent quantum efficiency and an action spectrum quite similar to that obtained for photooxidation of P700 which is consistent with the notion that they are associated with the reduction of X by P700.[35] Low-temperature electron spin resonance studies indicate the formation of an unpaired electron by the photo act.[36] The character of this spin resonance signal is quite similar to that observed with purified ferredoxin, but the ferredoxin can be removed and the signal persists.[37] Thus, another species of iron-sulfur protein, tightly bound to the chloroplast lamellae, may be X. It is assumed that X has a redox potential below -0.42 volts since it reduces ferredoxin. Several studies have been made with indicator dyes of very negative redox potential to estimate how low a redox potential is generated in Photosystem I. Since such dyes are extremely autooxidizable, the measurements must be done under anaerobic conditions with provision for removing any O_2 that arises from the Photosystem II splitting of water — usually with an excess of glucose and glucose oxidase. Various dipyridyl salts with a redox potential of -0.5 volts are reduced by illuminated chloroplasts.[38,39] Kok found that viologen dyes at -0.6 volts were reduced and he estimated a minimum potential of -0.7 volts for X.[40] Unlike the dyes used in these measurements, X as it resides in the spinach chloroplast is not autooxidizable. Some chloroplast preparations are able to catalyze a light-dependent oxygen consumption by the Mehler reaction as described by Figure 7.10. Here oxygen production is one-half of oxygen consumption. If catalase is added — $H_2O_2 \longrightarrow H_2O + \frac{1}{2}O_2$ — there will be no net change in oxygen, but one can measure isotope exchange if either the oxygen in the atmosphere or the water is labelled with O^{18}. With higher plant chloroplasts, Mehler reaction activity is dependent on either an added or a contaminating autooxidizable substance like viologen dye or a flavin to mediate the transfer of electrons from X to oxygen. Honeycutt has found a vigorous Mehler reaction in sucrose gradient washed preparations from blue-green algae suggesting that X may be autooxidizable in these plants.[41]

Parson has done some interesting kinetic studies which may describe properties of X in chromatophore preparations from photosynthetic bacteria.[42] An intense 20 nanosecond flash of light is able to oxidize P870 (the bacterial equivalent of P700) and cytochrome C_{422} (as evident from spectral bleaching of

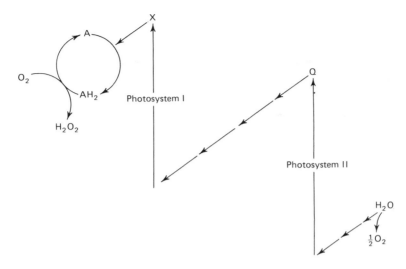

Figure 7.10   The reduction of an auto-oxidizable compound A by photosynthetic electron transport. In the absence of catalase activity,
this reaction gives a net consumption of oxygen.

these pigments) and these substances are presumed to be rather immediate
electron donors for the bacterial photo act. The photooxidation is followed by a
refractory period during which the system cannot use light. The recovery of light
utilizing ability (i.e., light-induced oxidation of P870 and cytochrome C_{422}) is
half complete in 60 microseconds at pH 7 and 80 microseconds at pH 8 and is
temperature dependent. The refractory period is perhaps due to accumulation of
reduced X. The X is estimated to be present in a one-to-one ratio with P870 and
its reoxidation is presumed to be relatively slow compared to its photo
reduction.

Yocum and San Pietro have reported on the isolation of a ferredoxin reducing
substance from spinach chloroplasts.[43],[44] They have isolated from chloroplasts
a substance that can be reduced by chloroplasts and oxidized by ferredoxin. The
isolation procedure, UV absorption spectrum, stability, and low molecular
weight (approximately 6,000) all suggest that this material is similar to the
cytochrome reducing substance isolated from algae and higher plants by Fujita
and Myers[45] and to the oxygen reducing substance isolated from blue-green
algae by Honeycutt.[41] The probable identity of these factors is further
supported by their common ability to react with an antibody that Regitz
et al. had prepared.[46] This antibody was shown to be a specific inhibitor of
Photosystem I and its inhibitory effect is nullified by preincubation with
ferredoxin reducing substance, cytochrome reducing substance, or oxygen
reducing substance.

7.6   LIGHT ABSORPTION FOR PHOTOSYSTEM I

The main light-collecting pigment in Photosystem I is chlorophyll a. The principal absorption bands are at 440 nm (blue) and 663 nm (red) when chlorophyll a is dissolved in an organic solvent. The red absorption band is shifted to higher wavelengths in dimerized and crystalline chlorophyll.[47] All fluorescence emission occurs at wavelengths above the red absorption band. The spectra in Figure 7.11 illustrate the above observations. The *in vivo* spectrum of

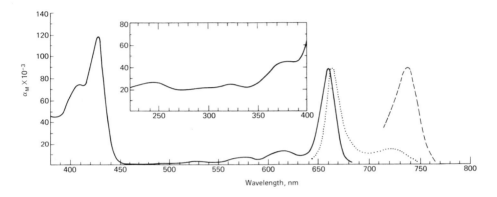

Figure 7.11   The absorption spectrum of chlorophyll a dissolved in ether is described by the solid line. The dotted line is the fluorescence emission spectrum. The dashed line is the absorption spectrum of microcrystals of chlorophyll. (From Kok, B., in *Plant Biochemistry*, J. Bonner and J. Varner, eds., Academic Press, N. Y., 1965, p. 909.)

chlorophyll is shifted toward the red as a result of protein-chromophore interaction. During the greening process in higher plants, the red peak of chlorophyll a made on protochlorophyll holochrome is first seen at 684 nm, and then it shifts to 673 nm where it remains. Careful examination of the chlorophyll a red band by derivative absorption (the first derivative of the absorption tends to magnify slight differences on the shoulders) or by low-temperature spectroscopy (to sharpen the bands) suggests that there may be several forms of chlorophyll a differing in their absorption peaks and perhaps their protein binding. *Scenedesmus* – a green alga – mutant #8 is said to lack a special chlorophyll 700 compared to wild type.[48] Figure 7.12 shows a careful spectroscopic comparison of the deficient mutant and the wild type. Multiple chlorophyll a fluorescence bands are observed at low temperature in some species.

The isolation of a pure chlorophyll-protein complex is a foreseeable goal. The isolation of protochlorophyll holochrome that turns green in the light was reported in the 1950's. This protein is big – its molecular weight is 500,000.

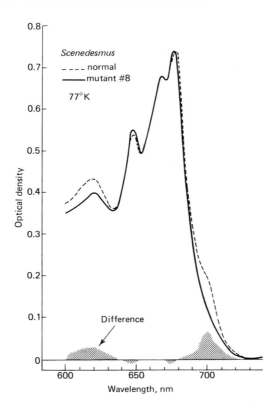

Figure 7.12   The low temperature absorption spectra of mutant number 8 of the green alga *Scenedesmus*, which lacks a functional Photosystem I (solid line), and of the wild type alga (dashed line). (From Kok, B., *Fluorescence Studies*, in "Photosynthetic Mechanisms of Green Plants," Natl. Res. Council, Misc. Publ. No. 1145, 45-55 (1963).)

Thornber et al. reported on solubilizing green proteins from chloroplasts with sodium dodecyl sulfate.[49,50] Two green fractions could be resolved by gel electrophoresis. These bands differ in amino acid and pigment content and in sedimentation velocity. These bands might be fragments from the two different photosystems. A very interesting water soluble chlorophyll-protein complex is obtained from cauliflower. This complex has a molecular weight of 78,000.[51] There are six chlorophyll a molecules for each chlorophyll b and the chlorophylls can be reversibly dissociated from the protein. This preparation is unique in that it is achieved without the use of detergents. Some chlorophyll-protein complexes are recognizable as *in vivo* artifacts or *in vitro* artifacts created by the detergents or organic solvents used in their isolation. Olsen has crystallized a bacteriochlorophyll-protein complex from *Chloropseudomonas ethylicum* and finds five chlorophylls per subunit of a protein with four identical subunits.[52] Such preparations should prove useful in describing the bulk form of chlorophyll in the plant at a molecular level. One would like to have a pure pigment-protein with a proper spectrum to better understand the light-harvesting function and through its physical and chemical characteristics the chloroplast architecture. It appears that there is not a high degree of geometrical regularity of chlorophyll-protein in the chloroplast as judged by ORD, CD, and X-ray scattering. The resonance transfer of photon energy may obviate the need for highly ordered pigment arrays. The photon can rapidly migrate through adjacent chlorophylls until it reaches a pigment with a long wavelength absorption maximum. The energy loss on entering this pigment traps the photon irreversibly and it must migrate to a pigment with a still longer wavelength maximum, or do chemical work, or be lost as heat or light.

7.7   P700, THE PRIMARY OXIDANT OF PHOTOSYSTEM I

Kok has suggested that the long wavelength chlorophylls could assist in funneling the quanta from the bulk collectors to the trap P700 where chemistry takes place. This mechanism is illustrated in Figure 7.13. In order to look for a

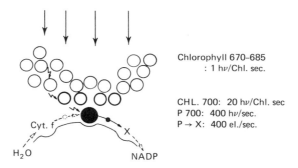

Chlorophyll 670–685
: 1 hν/Chl. sec.

CHL. 700: 20 hν/Chl. sec
P 700: 400 hν/sec.
P → X: 400 el./sec.

Figure 7.13   A two-step energy focusing mechanism for Photosystem I. Quanta are collected by the bulk chlorophyll molecules with absorption peaks at 670–685 nm. The quanta migrate irreversibly into chlorophylls with a peak at 700nm, then to P700 where the energy is expended in chemical oxidation. (From B. Kok, in *Plant Biochemistry*, J. Bonner and J. Varner, eds, Academic Press, N. Y., 1965, p. 909.)

trap where light energy is converted into chemical reaction, certain characteristics could be postulated. From the Emerson-Arnold experiments concerning the size of the photosynthetic unit (approximately 2,000 chlorophylls per O_2 or 1,000 chlorophylls per water split, or 500 chlorophylls per electron from water to NADP, or 250 chlorophylls per electron per photo act) it has been found that the number of traps must be less than 1% of the number of chlorophylls present.

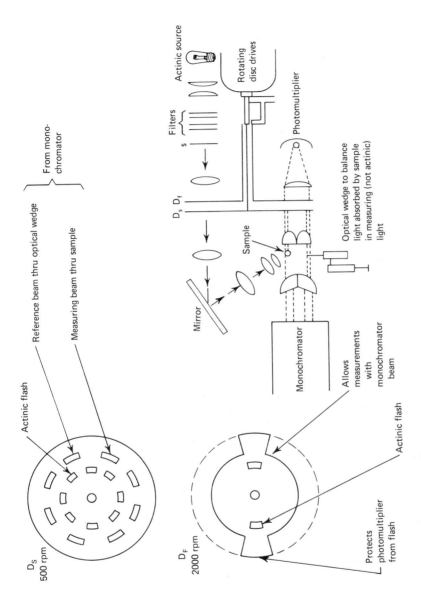

Figure 7.14 A sensitive spectrophotometer designed to measure light-induced changes in chloroplasts. [From Kok, B., *Plant Physiol.*, **34**, 184 (1959).]

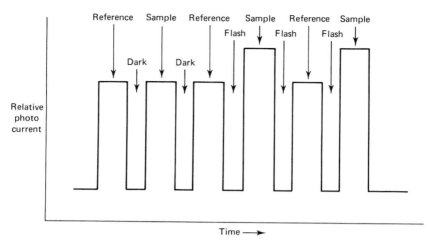

Figure 7.15   Changes in the light-absorbing capacity of chloroplasts after exposure to an intense flash of actinic light. [Redrawn from Kok, B., *Plant Physiol.*, **34**, 184 (1959).]

One needs an extremely good spectrophotometer to see a trap against a heavily pigmented background, and Kok built such a device to find P700.[53] A schematic diagram of this spectrophotometer is shown in Figure 7.14. A beam of low-intensity measuring light is obtained from a monochromator and divided into two beams. One beam passes through a sample of chloroplasts or algae and then falls on a photomultiplier. The parallel beam passes through an optical wedge that is adjusted to absorb exactly the same amount of light as the chloroplast sample, and then this standard reference beam, with the sample measuring beam, is focused on the photomultiplier. Using a single source for measuring light and for reference light makes the instrument self-correcting for fluctuations in intensity of light from the monochromator. Using a single photomultiplier makes the instrument self-correcting for fluctuations in sensitivity of the detecting system. If anything changes the absorbance of the chloroplast sample, the split-beam arrangement poises the photomultiplier to maximum sensitivity in seeing this change by quickly comparing it to a reference beam set at the same original intensity as the sample beam. The sample is then flashed from the side with actinic (action producing) light that causes photosynthesis. A fast rotating disc is used to give flashes of light on the sample and at the same time to cover the photomultiplier so that it is not blinded by the intense actinic light. Between flashes, a second rotating disc lets the photomultiplier look alternately at the reference beam and the measuring beam. If the actinic light causes a decrease in light absorption by the sample, say by the oxidation of a cytochrome when the measuring beam is being absorbed by the alpha band of a cytochrome in the photosynthetic chain, then the photomultiplier will see a little more light coming through the sample than from the reference beam. An oscilloscope measurement of the photomultiplier response would resemble the result shown in Figure 7.15. In effect, the machine is looking at the chloroplasts in the dark just after a flash to see if anything happened. The measurement is fast and repetitive, so that individual data can be fed into a computer of average transients (CAT) and assessed for significance at maximum sensitivity. All manner of electronic elaborations and variation in flash sequence can be arranged. By varying the wavelength of the monochromter, Kok obtained a spectrum of the pigment changed by illumination, and this is shown in Figure 7.16. Either ferricyanide or light will cause the bleaching of

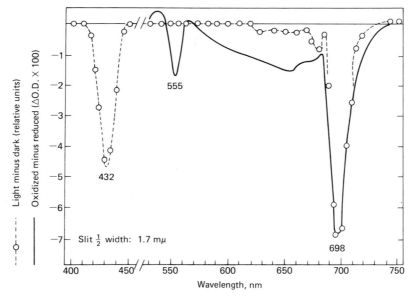

Figure 7.16   Light-minus-dark difference spectrum (dotted line) and oxidized-minus-reduced difference spectrum (solid line) of chloroplasts extracted with a water acetone mixture to remove much of the bulk chlorophyll. Both spectra show the P700 bleaching at 698 and 432 nm, and the oxidized-minus-reduced spectrum shows the alpha band of chtochrome f at 555 nm. [From Kok, B., *Biochim. Biophys. Acta*, 48, 527 (1961).]

pigment — both give a decrease in absorbance at 698 and 432 nm. In addition, ferricyanide reveals the alpha band of cytochrome f at 555 nm because of its chemical oxidation by ferricyanide. Obviously, light is causing the oxidation of a substance with a spectrum like chlorophyll but with the peaks shifted to longer wavelengths, making it an ideal trap for quanta. This trap, called P700, is normally reduced in the dark and oxidized by light.[54] By assuming that the P700 has the same extinction coefficient as chlorophyll a, which it resembles in absorption spectrum, one can estimate 1 P700/400 chlorophyll a molecules. This assumption has been confirmed and refined by experimental measurements.[55]

The magnitude of the measured change in absorbance on the usual scale is $\Delta A_{\lambda=700 \text{ nm}}$ = 0.010, if $A_{\lambda=675 \text{ nm}}$ = 4.000. Since P700 is oxidized by ferricyanide, its redox potential can be estimated by redox equilibrium measurements with a mixture of ferri- and ferrocyanide, giving an approximate E_O' = +0.43 volts.[56]

Assuming that P700 reduces X at an E_O' = -0.6 volts, one can calculate the necessary energy input from light to accomplish this electron transfer against the gradient of electrical potential. P700 at +0.43 volts to X at -0.6 volts is a span of 1.03 volts.

$$1.03 \text{ volts} \times \frac{96{,}494 \text{ joules}}{\text{volt}} \times \frac{1 \text{ calorie}}{4.18 \text{ joule}} \times \frac{1 \text{ Kcal}}{10^3 \text{ cal}} =$$
$$24 \text{ Kcal/mole of electrons moved}$$

compared to a mole of quanta at 700 nm

$$\frac{6.67 \times 10^{-27} \text{erg sec}}{700 \times 10^{-7} \text{cm}} \times \frac{3 \times 10^8 \text{ cm}}{\text{sec}} \times \frac{6.02 \times 10^{23} \text{quanta}}{\text{Einstein}} \times \frac{1 \text{ joule}}{10^7 \text{ergs}} \times$$
$$\frac{1 \text{ cal}}{4.18 \text{ joule}} \times \frac{1 \text{ Kcal}}{10^3 \text{ cal}} = 40 \text{ Kcal/Einstein at 700 nm}$$

So there is adequate energy in a quantum of red light to move an electron from the very poor reducing agent P700 to make a good reducing agent of E_O' = -0.6 volts or lower.

P700 is bleached with high photochemical efficiency – the quantum yield is 1. P700 is bleached instantaneously, as is proper for a photochemical reaction, while the half time for dark reduction is about 5 microseconds, which is proper for an enzymatic reaction. P700 is bleached at liquid nitrogen temperature (-150°C) as is expected for a photochemical reaction, but it is not re-reduced at this temperature.

On bleaching, P700 loses a single electron, and the oxidized form may be responsible for a free radical signal seen in photosynthetic tissue in response to light. A narrow band with fast kinetics appears in the ESR spectrum of chloroplasts on exposure to light.[57] The kinetics correlate well with the P700 absorption change, although there is a discrepancy between the number of spins and the amount of P700 oxidized. Bishop found a mutant of *Scenedesmus* that lacked the P700 absorption changes and the fast ESR signal.[58,59] Since this mutant was incapable of complete photosynthesis, it provided strong support for P700's role in the overall process. Similar mutants have been found in *Chlamydomonas* and *Rhodopseudomonas spheroides.*

In photosynthetic bacteria, there is a trap analogous to P700, but since the bulk bacteriochlorophyll spectrum is shifted far to the red, the trap is seen at 870 nm. This trap is generally similar to P700. In the bacteria there is an

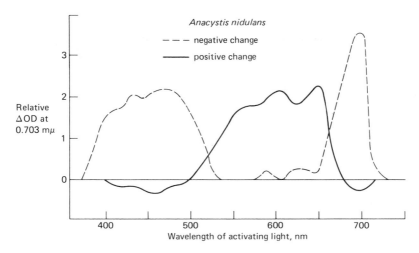

Figure 7.17   The effects of wavelength upon the redox state of P700 in whole cells of the blue-green alga *Anacystis*. The negative change which signals the photooxidation of P700 is measured without background and shows the reaction is driven by light absorbed by chlorophyll a. The positive change is measured against a background of dim red light to oxidize P700 and describes the abilities of other wavelengths to overcome the red light and cause reduction of P700. This action spectrum for the reduction of P700 corresponds to the absorption spectrum of the phycobiliproteins in this alga. [From Kok, B. and Gott, J., *Plant Physiol.*, **35**, 802 (1960).]

additional correlative evidence relating the ESR signal to P870. An ESR signal is induced by chemical oxidation of purified bacteriochlorophyll by iodine, and this signal is similar to the light-induced signal in intact cells.[60] The signal characteristics are altered in a similar way by using either deuterated bacteriochlorophyll or deuterated bacteria (i.e., grown in D_2O for enough generations to dilute out all of the hydrogen).[61]

Although several attempts have been made to isolate P700 from higher plant material, chloroplasts have proved refractory and blue-green algae have allowed modest progress.[62] Clayton has isolated the reaction center from *Rps. spheroides* by dispersing chromatophores with the detergent Triton X 100 and centrifuging on a sucrose gradient to get a band free of bulk chlorophyll and containing one micromole of P890/230 mg protein.[63] The protein is probably not homogenous, but this preparation represents real progress.

Kok has established the position of P700 relative to the two photosystems by varying the wavelength of the actinic light which affects P700 as shown in Figure 7.17. The blue-green alga *Anacystis nidulans* was used since the accessory pigment, phycocyanin, absorbs well between 500 and 650 nm in a way that is distinct from the red and blue absorption bands of chlorophyll a. In the experiment described by the graph one sees that irradiation of chlorophyll a by actinic light of the proper wavelength causes oxidation of P700. Irradiation of the phycocyanin to activate Photosystem II causes reduction of P700.

GENERAL REFERENCES

Bishop, I.N. "Photosynthesis and Electron Transport Systems of Green Plants," *Ann. Rev. Biochem.*, **40**, 197 (1971).

Gregory, R. P. F. *Biochemistry of Photosynthesis.* New York: Wiley-Interscience, 1971.

Rabinowitch, E., and Govindgee. *Photosynthesis.* New York: Wiley, 1969.

San Pietro, A. (ed.). "Photosynthesis," *Methods in Enzymology* **23**, Academic Press, New York, 1971.

8 PHOTOSYNTHESIS: ELECTRON TRANSPORT LINKING THE PHOTO ACTS, PHOTOSYSTEM II

Photosystem I consumes reducing power that must ultimately be traced back to water. This reducing power comes via the photo act of system II to an electron transport chain, which by exergonic reactions delivers the reducing power to Photosystem I. The participants in this chain — cytochromes, a copper protein, a quinone — are analogous to participants in the mitochondrial respiratory chain and so also are the modes of interaction. Recalling that P700 is oxidized by Photosystem I, one must ask what will reduce it to maintain the flow of electrons. Both plastocyanin and cytochrome f (a c-type cytochrome more accurately identified as cytochrome c_{554}) have been popular candidates as the reductant of P700. Since the redox potential of these two carriers is very similar, this is no help in assigning sequential order. Current evidence points to plastocyanin as the reductant of P700 in higher plants and a cytochrome as the donor in bacteria. Consider first the molecular properties of these two carriers and then the enzymatic evidence for their participation in photosynthesis.

139

8.1   PLASTOCYANIN

Plastocyanin is a copper protein that was first discovered by Katoh and Takamiya in *Chlorella*.[1,2] Plastocyanin has been isolated from higher plants and blue-green algae, but it has not been detected in photosynthetic bacteria. Plastocyanin is held in the chloroplast by moderately tight binding and accounts for about one-half of the copper in that structure. Plastocyanin is solubilized by sonication, extraction with concentrated salt solutions, or made readily soluble by prior extraction of the chloroplasts by heptane. There is one mole of plastocyanin per 300 to 500 chlorophylls in spinach. Plastocyanin is blue in the oxidized state and colorless in the reduced form. Absorption spectra are shown in Figure 8.1. Since the blue peak at about 600 nm is a bit low on a per mole basis and lies in a region of the spectrum where there is much chlorophyll absorption, it is difficult to observe plastocyanin in *in situ* spectroscopic studies. This copper protein is not autooxidizable, and the redox potential is +0.39 volts. The protein has a molecular weight of 10,000 to 20,000, depending on the species of origin, and contains two coppers which are the site of oxidation-reduction and which are responsible for the blue color. These coppers are not accessible to chelators. The copper in plastocyanin is said to show a negative Cotton effect, indicating asymmetry of the copper site, while ESR indicates both coppers are divalent and equal. Proton relaxation measurements indicate that the coppers are buried and that plastocyanin's catalytic function is not blocked by copper chelating agents.

Numerous experiments support a role for plastocyanin in photosynthetic electron transport. Extraction of plastocyanin from chloroplasts by sonication

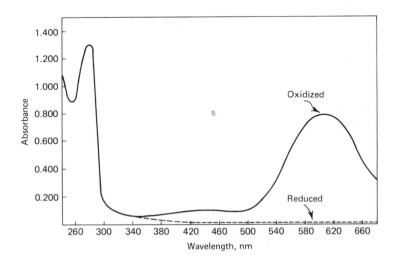

Figure 8.1   Absorption spectra of oxidized and reduced plastocyanin.

or detergent washing leads to a loss of Hill reaction activity. Readdition of purified plastocyanin gives a partial restoration, but the extraction procedure often damages other sites in the chain. A useful partial reaction of photosynthesis was discovered by Vernon, who found that chloroplasts whose oxygen photoevolution activity had been destroyed by heat or certain inhibitors could still photoreduce low-potential acceptors like NADP if the inhibited chloroplasts were supplied with reduced indophenol dye. Here the reduced dye is able to act as an electron donor in place of water, bypassing the inhibited sites that are presumably close to the oxygen-evolving step. This reaction between reduced dye and NADP is completely lost on removal of plastocyanin and can be restored almost completely on readdition of plastocyanin.

Several lines of experimentation indicate that plastocyanin is the immediate electron donor that reduces the photooxidized P700. Levine has provided some very attractive evidence in favor of a plastocyanin-to-P700 sequence by isolating and characterizing appropriate mutants of the alga *Chlamydomonas reinhardi*.[3]

Photosynthetic mutants are obtained by mutagenic treatment of algae which are facultative photoautotrophs — species that can grow in the light on CO_2 and in the dark on a fixed carbon source like glucose or acetate.[4] *Chlamydomonas* and *Scenedesmus* are two species that have been used successfully. After the mutagenic treatment, the surviving cells are grown out in the dark on a fixed carbon source and then screened for photosynthetic activity, either by autoradiographic identification of colonies that cannot fix CO_2 after exposure to $C^{14}O_2$ and light or by observing the plated colonies in red light with a far red filter since the mutants, which cannot use the absorbed light energy for photosynthesis, will emit the energy as far red fluorescence.

Gorman and Levine, using *Chlamydomonas* mutants grown on acetate, were able to characterize one mutant as lacking cytochrome f and one as lacking plastocyanin. Ac 206 is lacking cytochrome f as evidenced by the absence of the 552–554 nm alpha band in the difference spectrum. Ac 208 is lacking plastocyanin as evidenced by the failure to find this protein or even copper in the appropriate fractions. The mutants cannot photoreduce NADP with H_2O as an electron donor, but Ac 206, which lacks cytochrome f and retains plastocyanin, will do a partial reaction from reduced indophenol to NADP while Ac 208, lacking plastocyanin but containing cytochrome f, will not transfer electrons from reduced dye to NADP. These results are interpreted as shown in Figure 8.2.

Levine has been able to restore the activity of preparations from Ac 208 in the H_2O to NADP reaction by adding purified plastocyanin. Ac 206 activity could not be restored by adding cytochrome f.

When plastocyanin is removed from spinach chloroplasts, readdition of this copper protein will restore the lost ability to oxidize cytochrome f with Photosystem I light and to reduce NADP with reduced indophenol dye.[5] Similar reconstruction experiments have been achieved with chloroplasts from a variety of plant materials.[6,7,8] Plastocyanin *in situ* is not inhibited by an antibody to it

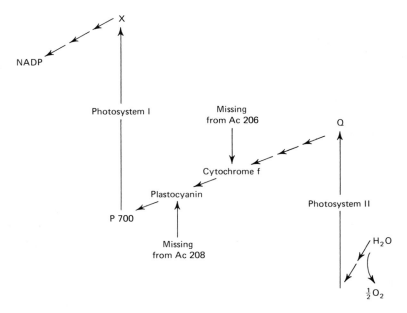

Figure 8.2   The specific loss of individual electron carriers through mutations of *Chlamydomonas reinhardi*.

unless the antibody is inserted into the chloroplast structure.[9] Plastocyanin
function is inhibited by polycations but only under conditions in which the
chloroplast membranes are spread open.[10]

8.2   CYTOCHROME "f"

Cytochrome f was discovered by Hill through the use of the low-dispersion hand
spectroscope like that instrument used by MacMunn almost a century ago when
he first described muscle cytochromes. The name cytochrome f is trivial but
distinctive for that c-type cytochrome associated with photosynthesis. Hill was
able to solubilize and purify cytochrome f from parsley.[11] Cytochrome f is not
autooxidizable and has a redox potential of +0.365 volts. Absorption spectra
typical of a cytochrome f are shown in Figure 8.3. As in the case of P700,
plastocyanin, and many other cytochromes in which there is a measurable
spectral change associated with oxidation or reduction, one uses a redox couple
of known potential which will spontaneously shift electrons to and from
cytochrome f in order to measure the redox potential. Ferricyanide and
ferrocyanide or various redox dyes are used as redox standards. One mixes some
known amounts of oxidized and reduced standard with a known amount of
cytochrome and measures the concentrations of the oxidized and reduced forms

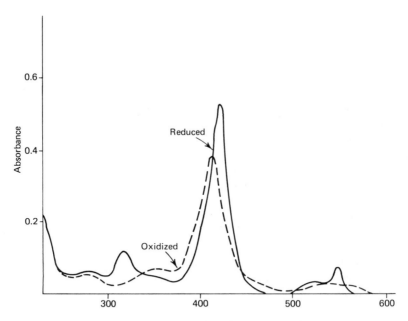

Figure 8.3   Absorption spectra of oxidized and reduced cytochrome f.

of both the standard and the cytochrome at equilibrium. This gives data for an equilibrium constant (K_e), which can then be used to calculate the difference in redox potential between the potential of the known standard and that of the cytochrome.

$$\Delta E_o' = \frac{RT}{nF} \ln K_e$$

If one knows the absolute value for the potential of the standard and the value of the difference in potential between the standard and the cytochrome, the potential of the cytochrome is obtained by simple arithmetic.

Having characterized a cytochrome f in parsley, Hill and others set out to find it in other photosynthetic tissue. A condensed compilation of the resulting data (Table 8.1) shows the general range of variability among analogous cytochromes from different sources.[12]

Cytochrome f can be released by gentle procedures from the photosynthetic membranes of several higher plants and algae and purified to homogeneity.[13,14] In photosynthetic bacteria, the analog of cytochrome f is called cytochrome c_2 and has been characterized in several genera as having an E_o' between +0.308 and +0.350 volts. Often the cytochromes of higher plants are called cytochrome c_{552} or cytochrome c_{554} rather than cytochrome f. All are c-type cytochromes with an alpha band between 550 and 555 nm and an E_o' between +0.30 and

TABLE 8.1   c-TYPE CYTOCHROMES FROM PHOTOSYNTHETIC ORGANISMS

Source	Absorption Peaks of Reduced Form (nm)		E'_O (V) pH 7	Molecular Weight
Parsley	554–555	421	+0.365	110,000
Euglena gracilis	552.5	416–417	+0.37	17,000
Chlamydomonas reinhardi	552.5	416.5	+0.37	12,000
Monostroma nitidum	552	416	+0.31	
Ulva	552	415	+0.30	
Porphyra tenera	553	417	+0.34	
Navicula	554	418	+0.34	13,000/heme
Anacystis nidulans	554	416.5	+0.35	23,000/heme
Rhodospirillum rubrum	550	415	+0.338	12,750

+0.37 volts. In the amino acid sequence extant for the cytochrome of *R. rubrum*, Kamen has documented an interesting internal homology suggesting partial gene duplication in the evolution of this protein and he has found some external homology to the respiratory chain cytochrome c.[15]

Not only is this type of cytochrome uniquely associated with photosynthetic tissue and absent from etiolated tissue, but Hill also found spectroscopic evidence for its involvement in photosynthesis. Cytochrome f is in the reduced form in the dark and becomes oxidized in the light. Duysens did the action spectra of cytochrome f oxidation and reduction in algae and found that Photosystem I light oxidized and Photosystem II light reduced this cyto-chrome.[16]

Chance has studied the temperature insensitive oxidation of a cytochrome in the photosynthetic bacterium *Chromatium*.[17] *Chromatium* probably lacks plastocyanin in any case, but one might think that the photochemical oxidation would be confined to P870 (analogous to P700 found in higher plants). Parsons looked at P870 and cytochrome changes with the Q-switch ruby laser as the illuminating source.[18] The following facts come from such measurements:

1. P870 is completely oxidized in 0.5 μseconds light at either 80°K or 298°K.

2. P870 reduction is complete in 2 μseconds, during which time a cytochrome band at 422 is oxidized. The P870 reduction does not occur at liquid nitrogen temperature, but the cytochrome oxidation does.

The room temperature data imply a cytochrome to P870 sequence, but the liquid nitrogen temperature data suggest that the cytochrome is either acting in parallel to P870 or is frozen onto P870 in a way that obviates the need for any molecular motion and allows the electron to pass right through the P870.

Hildredth used this laser method to look at cytochrome f oxidation in leaves and in isolated chloroplasts from spinach.[19] He found that cytochrome f oxidation is twice as fast in intact leaves as in chloroplasts, suggesting a structural dislocation during organelle isolation. The quantum requirement is 3 (instead of 1) for cytochrome oxidation in the intact leaf. Duysens had found a quantum requirement value of 1 with algae and one might expect high efficiency for a short span of electron transport. Avron reported a quantum requirement of 1 for chloroplast cytochrome f oxidation, so this process appears to be slower but more efficient in the isolated chloroplast. Perhaps there is a compensating absorbance change in the intact leaf that prevents measurement of all of the cytochrome f change. Hildredth found that the half-time for cytochrome f oxidation was 0.06 to 0.15 m seconds in the leaf, which is in the range of a very fast enzymatic reaction. He examined *Chlamydomonas* in this apparatus in order to measure P700 and the half time of oxidation of cytochrome f and of cytochrome b, which is thought to be further back in the sequence, i.e., closer to Photosystem II.[20] With these cells, Hildredth could detect a 30 μsecond lag in cytochrome f oxidation, hinting that something might be between cytochrome f and P700, but its turnover would be quite fast. Since by comparison there is a 300 μsecond lag between cytochromes b and f, there are surely carriers between these two. Several laboratories have found that cytochrome f photooxidation in higher plant chloroplasts is suppressed at low temperatures as would be expected for this plastocyanin-mediated reaction.[21,22]

One would hope that chloroplast depletion and reconstitution experiments might provide some help for this situation, but such experiments have not yet been possible. Higher plant chloroplasts bind cytochrome f tightly, so that this carrier has yet to be extracted under conditions which do not wreck other catalysts in the chloroplast. In algae the situation is more favorable. Katoh found that *Euglena* chloroplasts readily lost a cytochrome $c_{552} - E_o' = +0.3$ and apparently equivalent to cytochrome f.[23,24] The depleted chloroplasts could carry out a Hill reaction to high-potential electron acceptors like ferricyanide and indophenol dye but not to low potential acceptors like NADP or viologen dyes. Readdition of a catalytic amount of purified cytochrome f restores NADP reduction. In a system in which oxygen evolution is inhibited by the herbicide DCMU (dichlorophenyl-1, 1-dimethyl urea), the cytochrome is required for the transfer of electrons from reduced indophenol dye over Photosystem I to NADP. Net photoreduction of added cytochrome by the chloroplast preparation and photooxidation of reduced cytochrome were observed.

Pratt and Bishop have found several interesting cytochrome f-defective mutants of the alga *Scenedesmus*.[25] Fractionation of wild-type cells revealed a free or readily soluble form and a bound form of the cytochrome. Mutant number 50 lacks both the free and bound form of the cytochrome, but mutant 26 lacks only the bound form. Neither mutant can transfer electrons from water to NADP. Mutant 26 has the structural gene for cytochrome f production but is evidently defective in a critical binding site that holds cytochrome f in place and that is essential to its function. Both mutants show activity in the transfer of electrons from reduced dye to NADP and activity in cyclic phosphorylation, indicating that cytochrome f is not essential in these partial reactions.

Lightbody was able to remove both plastocyanin and cytochrome f from preparations of photosynthetic lamellae of the blue-green alga *Anabaena variabilis*. The extraction irreversibly inactivated Photosystem II, but Photosystem I activity, i.e., reduced dye to NADP, could be measured on addition of either plastocyanin or cytochrome f to the depleted lamellae. Plastocyanin was slightly more efficient in restoration but otherwise appeared interchangeable with the cytochrome f.[26] Trebst has devised three procedures to remove plastocyanin from spinach chloroplasts, and in each cytochrome f could replace plastocyanin in restoring Photosystem I activity.[27]

If the Levine mutant evidence is correct, one would expect no restoration of Photosystem I by cytochrome f unless plastocyanin were present. These two carriers, however, are so similar in redox potential that they may be freely interchangeable in a reconstituted system.

8.3   b-TYPE CYTOCHROMES

In addition to the c-type cytochrome, Bendall and Hill observed a cytochrome of the b-type in photosynthetic tissue.[28] They found that while the cytochrome f becomes reduced in the dark and oxidized in the light, the cytochrome b shows the reverse behavior — it is oxidized in the dark and reduced in the light. The E_o' of this cytochrome was estimated at 0.0 volts. These observations led Hill to the first formulation of the Z scheme. A general difficulty with the study of b cytochromes is their insolubility. Nevertheless, one b cytochrome has been purified from higher plant chloroplasts.[29] Additional isolation and characterization of b cytochromes may clarify their role, which is poorly understood because of the limitation that these catalysts have only been studied by *in situ* spectroscopy. It now appears that there are two b-type cytochromes in higher plant chloroplasts. They are distinguished by the different positions of their alpha bands and by differences in redox potential.[30]

Cytochrome b_6: Alpha band at 563 nm, E_o' = 0.0 volts, autooxidizable, not reduced by ascorbate, cytochrome b_6/cytochrome f = 1.6 (this is the original Hill and Bendall cytochrome b). Bendall has recently found that this can be resolved at low temperature into two alpha bands at 559 and 563 nm, both of which have an E_o' = 0.0 volts.

Cytochrome b_{559}: Alpha band at 559 nm, E_o' = +0.05 possibly to as high as +0.37, not autooxidizable and readily reduced by ascorbate, cytochrome b_{559}/cytochrome f = 2.

A species survey is given in Table 8.2. The relative positions and importance of these b-type cytochromes in the electron transport chain has been probed by kinetic studies but is still uncertain.[31] Hind and Olson found b_6 was reduced by Photosystem I light and the kinetics of this reduction showed no correlation

TABLE 8.2   b-TYPE CYTOCHROMES FROM PHOTOSYNTHETIC ORGANISMS

| | Reduced | | |
	Alpha (nm)	Gamma (nm)	E_o' (V)
Monostroma	562	428	+0.13
R. rubrum	560	430	0.0
R. vannielli	563	423	
Chlamydomonas b	559		>+0.05
Chlamydomonas b_6	563		>+0.05
Euglena	561	432	
Higher plant b	559		>+0.05
Higher plant b_6	563	429	0.0

with cytochrome f changes. Cytochrome b_{559} is oxidized by Photosystem I and reduced by Photosystem II.[32] These data might be rationalized as shown in Figure 8.4. Cramer and Butler showed b_{559} reduction in Photosystem II light

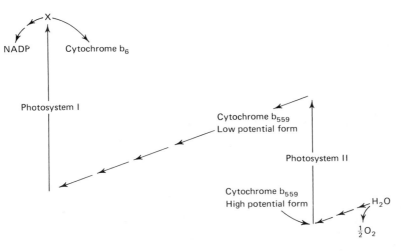

Figure 8.4   Localization of "b" type cytochromes in the photosynthetic electron-transfer sequence.

with its oxidation by Photosystem I.[33] The rates of the reduction and oxidation of the b cytochromes are slow, and b_6 vs. b_{559} rates are not clearly different; hence they are hard to distinguish.[34] Knaff and Arnon found that cytochrome b_{559} is photooxidized by Photosystem II light at liquid nitrogen temperature.[35] This led to the postulate that cytochrome b_{559} is the primary electron donor to Photosystem II. The basic observation has been confirmed in several laboratories, but the interpretation is very much a matter of dispute.[36,37,38] There are suggestions that cytochrome b_{559} is not on the main pathway of electron transport, but that it is a participant in a cycle of electron flow around Photosystem II or that only a part of this cytochrome (perhaps an aberrant part) is on the oxidizing side and the rest is on the reducing side of Photosystem II. Cytochrome b_{559} can exist in different states in the chloroplast and these states show differences in redox potential which might allow the cytochrome to function on either side of the photo act.[39] Levine has reported briefly on mutants of *Chlamydomonas* which support a sequence in which cytochrome b_{559} serves as an electron donor for a reaction that leads to reduction of cytochrome f.[40]

8.4   PLASTOQUINONE

The participation of lipid as an essential component in photosynthesis was first indicated by experiments in which the extraction of lyophilized chloroplasts with a non-polar hydrocarbon solvent removed Hill reaction activity.[41] When the extracted lipid was dried back onto the chloroplast, the activity was restored. Plastoquinone (PQ) was identified in photosynthetic tissue and localized in the chloroplast.[42] PQ was then shown to be the essential component removed by hydrocarbon solvent extraction.[43] The structure and absorption spectra of plastoquinone are shown in Figure 8.5. On closer examination a

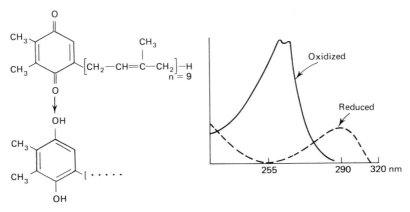

Figure 8.5   Structures and absorption spectra of oxidized and reduced plastoquinone.

variety of PQ analogs have been found and characterized. A PQ A with a C_{20} side chain has been isolated from *Aesculus*. A series of six isomers of PQ B and PQ C occurs in all higher plants and all algae with the exception of *Anacystis*, which may have a special replacement for PQ C.[44,45,46] The plastoquinone B series arises by esterification of a fatty acid to PQ C[47] (Fig. 8.6). The six isomers of PQ C are believed to be the result of hydration of a double bond at the second or some other more distal isoprene unit from the ring.

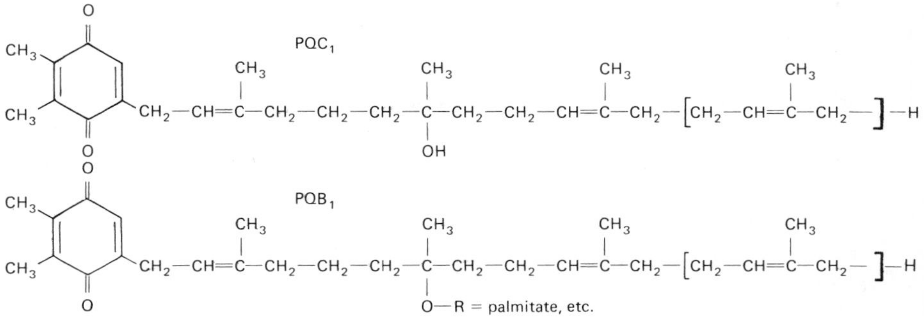

PQC$_1$

$$CH_2{-}CH{=}C(CH_3){-}CH_2{-}CH_2{-}CH_2{-}C(CH_3)(OH){-}CH_2{-}CH_2{-}CH{=}C(CH_3){-}CH_2{-}[CH_2{-}CH{=}C(CH_3){-}CH_2{-}]{-}H$$

PQB$_1$

$$CH_2{-}CH{=}C(CH_3){-}CH_2{-}CH_2{-}CH_2{-}C(CH_3)(O{-}R){-}CH_2{-}CH_2{-}CH{=}C(CH_3){-}CH_2{-}[CH_2{-}CH{=}C(CH_3){-}CH_2{-}]{-}H$$

O—R = palmitate, etc.

Figure 8.6   The structures of plastoquinones B and C.

In addition to PQ, higher plant chloroplasts contain tocopherols and tocopherolquinones.[48] One survey indicated a ratio of alpha tocopherolquinone to PQ of 0.4 to 2 in a variety of higher plants, with 95% of the alpha TQ in the chloroplast. Although such a concentration and location are suggestive of a role in photosynthesis, no conclusive evidence points to a function for the tocopherols in photosynthesis.

In the photosynthetic bacteria, PQ is replaced by CoQ in *R. rubrum*, by menaquinone in *Chloropseudomonas ethylicum*, or by chlorobium quinone in *Chlorobium thiosulfaticum*.[49,50]

PQ is abundant – 1 PQ A per 10 chloroplylls. Although PQ B and C are present in lower concentrations, they are still more abundant than cytochromes. PQ reduction *in vivo* has been detected spectroscopically in light–dark difference measurements. In isolated chloroplasts, illumination causes reduction of from 10 to 50% of the PQ A present, suggesting that the total pool need not participate in electron transport. PQ A reduction is better in Photosystem II light, while its oxidation is favored in Photosystem I light.

The redox potential of PQ is estimated at around 0.0 volts; therefore, like cytochrome b it is assigned a position near Photosystem II. Supporting evidence for this location of PQ A in the chain comes from extraction experiments. Although removal of PQ A prevents the flow of electrons from H_2O to NADP, it does not hinder the light-dependent movement of electrons from reduced indophenol to NADP.[51,52]

The relative positions of PQ and cytochrome b_{559} are not clear. The variable redox potential for cytochrome b precludes use of that property as a guide to whether PQ is reduced or oxidized by this cytochrome. The inhibitor 2,5 dibromo-3-methyl-6 isopropyl-p-benzoquinone shows some specificity in blocking the PQ dependent activities of chloroplasts, and this inhibition is reversed by addition of exogenous plastoquinone. Since this inhibitor blocks the reduction of cytochrome f by Photosystem II light and the oxidation of cytochrome b_{559} by Photosystem I light, plastoquinone is positioned between the two cytochromes.[53]

Crane has exhaustively extracted dry chloroplasts with heptane to remove both PQ A and PQ C.[54] The morphology of the membrane structures was little changed by the extraction except for a lessening of globular surface structures seen with the electron microscope. These preparations showed a need for both PQ A and PQ C in a ratio of their natural occurrence to give maximum restoration of Hill activity.

It was noted earlier that light induces the appearance of distinct electron spin resonance signals in photosynthetic tissue.[55] One is the fast, narrow band associated with P700. Another is a broad band with very slow kinetics which is attributed to plastoquinone. The signal diminishes on PQ extraction and is restored on PQ readdition, and controlled oxidation of reduced PQ in a model reaction produces a free radical with a spin resonance signal like the one seen in illuminated chloroplasts.[56] If the attribution of the electron spin resonance signal to PQ is correct, it does not prove that PQ is being oxidized or reduced via one electron steps exclusively during its photosynthetic function. However, a single electron mechanism is very attractive since one has single electron transfers via cytochromes and most likely by Photosystem II. The sluggish kinetics may be attributable to the large pool size of PQ.

8.5   PHOTOSYSTEM II

The chemical identity of redox participants beyond plastoquinone is obscure. Chlorophyll b (or phycobiliprotein in red or blue-green algae, or a carotenoid in brown algae) harvests short wavelength quanta and feeds them into Photosystem II. Chlorophyll a is present in Photosystem II and some special, unknown form of chlorophyll a presumably acts as a trap in a fashion analogous to P700 in Photosystem I. Kok has tried to estimate the redox potential of the reducing agent produced by photochemical activation of electrons in Photosystem II.[57] He used chloroplasts prepared from mutant #8 of *Scenedesmus*, which is either lacking or defective in the function of P700, so that Photosystem I is not operative. A series of oxidants were tested to find the lowest (most negative) E_o' oxidant that Photosystem II could reduce in the light. Electron acceptors of E_o' greater than +0.2 volts were easily reduced, but at +0.18 volts the acceptor was reduced with difficulty. Thus the E_o' of the reductant generated by Photosystem II is about +0.18 volts. Obviously, the redox potential relations around cytochrome b_{554}, PQ, and Q — the postulated terminus of Photosystem II — are fuzzy and these E_o' are not known with the same precision as those of carriers that have been solubilized and purified. The identity of the substance Q has been elusive and no help comes from the tendency to call it Y, Q, or E, depending on which author and what type of measurement is being made. A spectroscopic change at 550 nm is observed when chloroplasts are illuminated with light of those wavelengths absorbed by Photosystem II.[58] This change is known to be caused by the reduction of an electron carrier and, since it occurs at liquid nitrogen temperature, this compound might be the primary reductant of Photosystem II.

8.6   PHOTOSYSTEM II FLUORESCENCE

One experimental approach to the identity of Q comes from studies of chloroplast fluorescence. Govindjee first showed the relation of fluorescence yield to the two light reactions.[59] When chloroplasts are illuminated, especially in the absence of an electron acceptor, some of the absorbed light energy is lost by emission of light. Govindjee found that light absorbed by Photosystem II was much more effective in causing fluorescence than light absorbed by Photosystem I. In addition, fluorescence caused by exciting Photosystem II was diminished by exciting Photosystem I at the same time. Duysens and Sweer refined this observation and, noting that the fluorescence emission did not begin immediately when the light went on, they postulated the existence of Q — a quencher substance that is reduced by Photosystem II. During reduction of Q, since the absorbed quanta are moving electrons, the energy is used and fluorescence is quenched.[60] When Q is reduced, the absorbed quantum of energy is then emitted as light, i.e., fluorescence is no longer quenched. Reoxidation of Q by Photosystem I regenerates the quenching ability by allowing electrons to flow.

A comparison of fluorescence yield and electron transport helps to emphasize the inverse relation of chemical accomplishment to light emission. Data from such a comparison are shown in Figure 8.7. In this experiment chloroplasts are illuminated with blue light to excite the chlorophyll 420 band, and the fluorescence emitted at 700 nm and above is measured as a function of the intensity of the exciting light. In the curve labelled control, fluorescence is high since there is no electron acceptor and the chloroplast dissipates the absorbed energy by emitting quanta. If ferricyanide is added, fluorescence decreases since the absorbed energy can be used to push electrons to ferricyanide. If ferricyanide plus the uncoupler NH_4Cl is added, fluorescence is further decreased since energy that might be used in ATP synthesis, if only ADP and

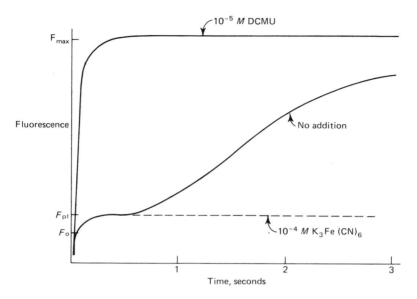

Figure 8.7   Fluorescence rise curves of chloroplasts as affected by the
inhibitor DCMU and by the electron acceptor potassium
ferricyanide. [From Forbush, B. and Kok, B., *Biochim.
Biophys. Acta*, **162**, 243 (1968).]

inorganic phosphate were present, can now be dissipated as heat in the presence
of the uncoupler, where previously it had to be dissipated as light. If you
consider the rate of dye reduction as a function of light intensity, you note that
the Hill reaction activity, or capacity to move electrons, is saturated with light at
an intensity at which the ferricyanide fluorescence curves inflect. This means
that when the rate of electron movement becomes limiting, more fluorescence
will appear to dissipate the excess absorbed energy.

One might guess that there are intermediate pools of electron carriers
adjacent to Photosystem II – perhaps Q and PQ – which, to the extent that
they are oxidized, would serve as sinks for an appreciable number of electrons
when the light is turned on. A very sensitive method is needed to measure this
amount of oxidant (usually these electron carriers are measured through the
catalytic turnover of the pools rather than the absolute pool size of the catalyst).
In the absence of an electron acceptor, one might find a miniscule oxygen
evolution and a low (submaximal) fluorescence yield immediately after the light
went on (while the endogenous pools fill with electrons); then oxygen evolution
will cease and fluorescence will go to a maximum as soon as the pools are full.
The kinetics of the fluorescence appearance, i.e., rise time, can be measured with
great sensitivity and shows a biphasic response. When Kok used a subsaturating
intensity of actinic light on isolated chloroplasts (Fig. 8.8), he saw an
instantaneous rise to F_o, then a plateau, then an additional rise in fluores-

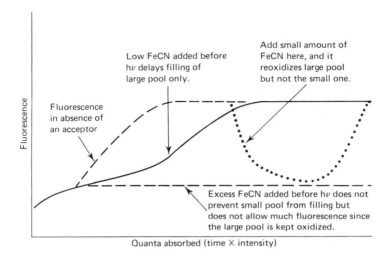

Figure 8.8   Changes in the appearance of fluorescence of chloroplasts which
                are the result of allowing electron transport to occur. [Redrawn
                from Malkin, S. and Kok, B., *Biochim. Biophys. Acta*, 126, 413
                (1966).]

cence.[61] If ferricyanide is added, the slow second phase is eliminated since the
electrons are drained off. If DCMU, a specific inhibitor of electron transport in
the region of Photosystem II, is added, maximum fluorescence is reached
immediately.

These two kinds of rise curves indicate the rapid filling of a small pool and
the slower filling of a large pool of oxidant. It is worth noting that addition of
the strong reductant, dithionite, prior to illumination causes an immediate rise in
fluorescence to the maximum, presumably by filling the pools by chemical
reduction before the light is turned on. DCMU causes an immediate rise in
fluorescence presumably by preventing the pools from draining.[62] Kok has
titrated the missing fluorescence when an electron acceptor is added vs. the
amount of acceptor reduced.[63,64] He estimates 1 carrier/1250 chlorophylls in
the small primary pool and 1 carrier/70 chlorophylls in the large secondary pool
(the large pool is called A or P in contrast to the small pool called Q or E).

An independent confirmation of this situation has come from the work of
Joliot, who developed an extremely sensitive oxygen electrode with a very fast
response time.[65,66] Joliot could measure the oxygen gush when *Chlorella* or
chloroplasts are illuminated in the absence of a Hill oxidant. This oxygen release
has the same kinetics as the fluorescence rise, and estimates of the size of the
endogenous electron acceptor pools agree rather well with Kok's estimates from
fluorescence measurements. Joliot estimates 1 carrier/500 chlorophylls in
Chlorella in the fast pool and 1/50 chlorophylls in the slow pool. By using
flashing light, Joliot found a minimum flash that would not give oxygen

evolution but would potentiate oxygen evolution by a subsequent flash. From this observation he postulates a photoactivation of the system prior to electron flow.

	Pool Size	Spinach Chloroplasts Kok	Chlorella Joliot
Q = E = Y	small, fast pool	1/1250	1/500
P = A	large, slow pool	1/700	1/50

Therefore, the kinetic evidence indicates that light absorbed by Photosystem II activates the mechanism and then reduces Q, from which electrons then flow to P. Joliot and Kok have indicated that quanta absorbed by Photosystem II and not used in the Photosystem II traps do not go on to drive the reactions of Photosystem I.[64,67]

The phenomena of delayed light emission refers to luminescence or delayed fluorescence emission of photons from chloroplasts long after the actinic light is turned off. The energy involved in this process is very slight, but it is an indication of a slight chemical reaction releasing energy which excites chlorophyll. Mutant studies indicate that delayed light is a property of Photosystem II and the preillumination conditions needed to get delayed light suggest an activation step similar to the one postulated by Joliot for oxygen emission.[68,69]

Kelley and Sauer have made a very interesting study of the unit size of the two photosystems in spinach chloroplasts.[70] The principle of the measurements is the same as that used by Emerson and Arnold in measuring the carbon dioxide fixing unit in intact algae. Kelley and Sauer measured the reduction of ferricyanide or indophenol dye and assumed that they are reduced only by Photosystem II. They calculated one reaction center for Photosystem II per 55 chlorophylls.

8.7   MANGANESE AND THE SPLITTING OF WATER

Manganese has been recognized as an essential mineral nutrient for acquisition of photosynthetic competence in higher plants, but it is not required in bacterial photosynthesis in which there is no Photosystem II. Nutritional deprivation of manganese results in chloroplasts unable to evolve oxygen but with residual Photosystem I activity. Reversal of manganese deficiency by addition of manganous salt to a deficient culture of *Anacystis* reveals a requirement for light but not a requirement for carbon dioxide fixation in restoring photosynthetic ability.[71,72] This light requirement is met through Photosystem II and may be because of the oxidation of the manganous ion to a higher valence state after it has been bound into the structure in a spontaneous dark reaction.[73] The photoactivation involves two distinct photo acts.[74] Manganous ion is removed from the isolated chloroplast with consequent loss of Photosystem II function by washing in concentrated Tris buffer. There is a partial reversal of this depletion of the cell-free preparation under proper conditions. Manganese binds to the chloroplast at 2 kinds of sites and Photosystem II contains 3 manganese atoms per 200 chlorophylls.[75,76]

Chloride, like manganese, plays some role in the vicinity of Photosystem II and affords the added experimental advantage of easily reversible removal from the isolated chloroplast.[77,78] Although the evidence is very tentative, the current suggestions would organize this information about Photosystem II as follows:

$$H_2O \xrightarrow[\text{dependent}]{Cl^-} \ ? \xrightarrow[\text{dependent}]{Mn^{+2}} \ ? \xrightarrow{\text{Photosystem II}} Q \xrightarrow[\text{DCMU}]{} A \longrightarrow$$

9 | PHOTOSYNTHESIS: PHOTO-PHOSPHORYLATION, INHIBITORS, PARTIAL REACTIONS, STRUCTURAL UNITS

In addition to reduced NADP, the dark reactions in the Calvin cycle of carbon dioxide fixation require ATP. Photosynthetic phosphorylation by isolated chloroplasts was discovered by Arnon in 1954 and meets the need for ATP which is explicit in the Calvin cycle of carbon dioxide fixation. Photophosphorylation is an activity of isolated chloroplasts that can be measured under two operationally distinct sets of circumstances.

9.1 STOICHIOMETRIC PHOSPHORYLATION

Stoichiometric photophosphorylation is quantitatively related to the flow of electrons from water to an appropriate electron acceptor such as NADP or ferricyanide. The process can be described by the equation

$$H_2O + NADP + nPi + NADP \xrightarrow[\text{chloroplasts}]{\text{light}} \tfrac{1}{2} O_2 + NADPH + H^+ + nATP$$

where n, the number of moles of ATP formed, is certainly 1 and possibly 2. Although the P/O ratio for NADH oxidation by tightly coupled mitochondria is conceded to be 3, the P/O ratio, or better the P/e_2 ratio — since the amount of Hill oxidant reduced is usually what is measured — is clearly 1 when no correction is made for the amount of electron transport in the absence of ADP. If one assumes that the electron transport seen in the absence of phosphate acceptor remains uncoupled to phosphorylation in the presence of Pi, ADP, and Mg^{+2}, then this inherently uncoupled rate should be subtracted from the higher measured rate when ATP synthesis occurs and the resulting P/e_2 ratio equals 2. Good has found that a variety of inhibitor effects support the $P/e_2 = 2$ figure, but this evidence is still indirect and contested.[2] An important point is the stimulation in rate of electron transport observed when ADP, Pi, and Mg^{+2} are added, i.e., an acceptor effect similar to "respiratory control" in mitochondria. Presumably, electrons will not flow if the energy released in oxidation cannot be used for ATP synthesis or otherwise be dissipated. A number of compounds such as amines, CCCP, and even dinitrophenol stimulate electron flow to the same extent as ADP, Pi, and Mg^{+2} but inhibit ATP formation — these compounds are "uncouplers" in that they dissociate electron transport from ATP synthesis and presumably cause dissipation of the energy of oxidation. Uncouplers are to be distinguished from electron transport inhibitors and from energy transfer inhibitors. This distinction is illustrated by the diagram in Figure 9.1.

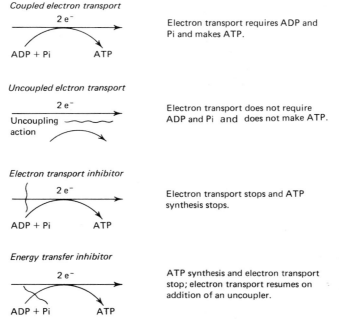

Coupled electron transport

Electron transport requires ADP and Pi and makes ATP.

Uncoupled elctron transport

Electron transport does not require ADP and Pi and does not make ATP.

Electron transport inhibitor

Electron transport stops and ATP synthesis stops.

Energy transfer inhibitor

ATP synthesis and electron transport stop; electron transport resumes on addition of an uncoupler.

Figure 9.1   A diagrammatic representation of the distinctions among inhibitory effects on chloroplast electron transport and phosphorylation. (From Good, N., unpublished.)

NADP, ferricyanide, and various quinones act as electron acceptors for stoichiometric phosphorylation as do FMN and menadione, although the latter two are autooxidizable and the reduced form does not accumulate. The phosphorylation site(s) is presumably located along the electron transport chain between or adjacent to the photosystems.

There have been several attempts at locating the site of stoichiometric phosphorylation between the photosystems. Trebst reported that reduction of NADP by reduced indophenol or diaminodurene via Photosystem I is stimulated by ADP + Pi or by uncouplers. This suggests that there is a phosphorylation site between the donor and Photosystem I.[3,4] Some confusion may be due to multiple sites of electron donation to the chain. Evidence will be noted later that donor systems act at several sites depending on donor concentration and other factors — one of these sites may pass through a phosphorylation step. Trebst has also developed evidence that there may be a phosphorylation site between the site of water splitting and Photosystem II, but this must remain tentative until more convergent evidence appears.[5] There has been no evidence to indicate a phosphorylation site between Photosystem I and NADP.

9.2   CYCLIC PHOSPHORYLATION

Cyclic photophosphorylation occurs when chloroplasts are illuminated in the presence of catalytic amounts of certain redox compounds such as pyocyanin or phenazine methosulfate (PMS). Huge amounts of ATP are synthesized without significant net consumption or production of oxygen. There is no significant O^{18} exchange from the water into the atmosphere or vice versa during the cyclic phosphorylation. Therefore, it is not a case of oxygen consumption equalled by oxygen production.[6] Cyclic phosphorylation is not inhibited by concentrations of DCMU that block all Hill activity, and this type of phosphorylation seems most efficiently driven by Photosystem I light. The current assumption is that electrons are donated to the cyclic cofactor by X or something near it at the top of Photosystem I and then returned to the chain in a cyclic fashion. The site of reoxidation of the cyclic cofactor is unknown. Extraction experiments indicate a PQ requirement for cyclic phosphorylation, but this may be structural rather than a chemical participation.[7] Inhibition of cyclic phosphorylation by the antibody to plastocyanin and by polycations indicates that plastocyanin is a participant in this cyclic electron transport pathway.[8,9] Pratt and Bishop have characterized a mutant of *Scenedesmus* as lacking cytochrome f.[10] Chloroplasts prepared from this mutant are active in cyclic phosphorylation. Cramer and Butler developed spectroscopic evidence for the participation of cytochrome b_6 in cyclic phosphorylation — b_6 turnover was stimulated by illumination plus PMS, and rate effects from ADP, Pi, and from uncoupling were seen.[11] These data suggest at least two sites of phosphorylation: one associated with cyclic, the other with non-cyclic phosphorylation, as envisioned in Figure 9.2. There is a differential sensitivity of cyclic vs. stoichiometric phosphorylation in blue-green algal preparations.[12] In these experiments, stoichiometric phosphorylation is much more labile to mechanical abrasion than either the Hill reaction or cyclic phosphorylation. Black has argued for multiple phosphorylation sites from his finding that exposure to heptane entirely destroys the chloroplast activity in stoichiometric phosphorylation and inhibits cyclic phosphorylation by 50%.[13,14] This is taken to mean that there are two sites in the cyclic pathway, one of which is used by the stoichiometric path; this latter site is sensitive to non-polar organic solvent destruction.

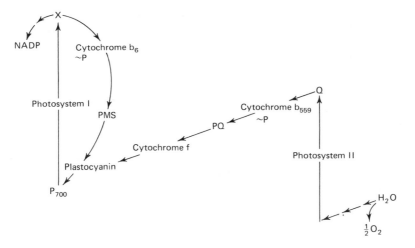

Figure 9.2   Suggested location of the phosphorylating sites coupled to photosynthetic electron transport.

An issue of considerable interest is whether cyclic phosphorylation is an *in vitro* artifact or something of physiological interest. The Calvin cycle predicts a requirement of 2 NADPH and 3 ATP per CO_2 fixed, and in tropical grasses the ATP requirement is still higher. If one accepts a $P/e_2 = 1$ for *in vivo* stoichiometric phosphorylation, additional ATP must come from either cyclic phosphorylation or some other method of utilizing NADPH. In any case, it would be nice to have cyclic phosphorylation to serve the many other ATP requirements of the cell.

Isolated chloroplasts require addition of some redox catalyst if cyclic phosphorylation is to be achieved. The requirement is non-specific — many artificial dyes as well as natural products, e.g., FMN, will work. Presumably, the "natural" carrier is washed out of the chloroplasts and it has proven difficult to identify a specific leaf constituent as *the* coenzyme of cyclic phosphorylation.[15,16,17]

An interesting approach to this problem has been to attempt whole cell experiments to test for cyclic phosphorylation *in vivo*. Forti poisoned leaf discs with DCMU in order to eliminate carbon dioxide fixation and found a net increase in ATP on shifting the discs from dark to light.[18] Hoch devised an extremely sensitive method for measuring light-induced passage of electrons through the cytochromes in intact algae. Again, although DCMU could poison the net flow of electrons from water, a rapid movement of electrons through a b-type cytochrome was observed.[19] Light-stimulated P^{32} incorporation into DCMU-poisoned algae and light-induced glucose uptake by algae, presumably an ATP dependent process, has been used as a measure of DCMU-insensitive (cyclic?) phosphorylation.[20,21] The failure of DCMU to prevent light-dependent movement in some motile algae is construed as evidence of cyclic phosphorylation. Thus there are many straws in the wind indicating an *in vivo* cyclic phosphorylation, but the current evidence is not yet conclusive.

9.3   PHOSPHORYLATION MECHANISMS

Regardless of the uncertainty about multiple sites of phosphorylation and multiple electron transport pathways serving these sites, one can still ask how does photophosphorylation occur. The two currently popular alternatives for a mechanism of photophosphorylation are the "chemical" and the "chemiosmotic" theories. The chemical theory is described by the series of equations:

$$AH_2 + C \longrightarrow AH_2-C$$

$$AH_2-C + B \longrightarrow A \sim C + BH_2$$

$$A \sim C + Pi \longrightarrow A + C \sim P\,(or\,A \sim P + C)$$

$$C \sim P\,(or\,A \sim P) + ADP \longrightarrow C\,(or\,A) + ATP$$

$$overall\ AH_2 + B + ADP + Pi \longrightarrow A + BH_2 + ATP$$

This scheme is based on an analogy to the substrate level phosphorylation catalyzed by glyceraldehyde-3-phosphate dehydrogenase. One might hope to isolate a phosphorylated intermediate $-C \sim P$ or $A \sim P$ — generated by electron transport activity, but very strenuous efforts to isolate a P^{32}-labelled intermediate have failed. One may suppose that $C \sim P$ is a very unstable or non-covalent association.

The chemiosmotic theory avoids this difficulty by postulating that the energy for the ATP synthesis comes from formation of a proton gradient.[22,23] Proton gradients are generated by both chloroplasts and mitochondria carrying out electron transport with attendant phosphorylation.

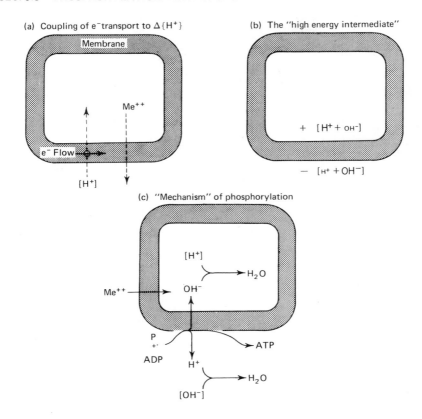

Figure 9.3   A diagrammatic representation of the chemiosmotic model for ATP synthesis. At stage (a) electron transport drives the accumulation of protons in a membrane enclosed space. At stage (b) the accumulated protons and hydroxyl ions on the different sides of the membrane are the "high energy intermediate" or chemical potential which, when discharged, drives the synthesis of ATP. [From Jagendorf, A. T., *Fed. Proc.*, **26**, 1361 (1967).]

The chemiosmotic mechanism can be represented as shown in Figure 9.3. The hypothesis demands an enclosed volume bounded by a membrane that has a low passive permeability to all ions. One can imagine the proton gradient being formed as follows. Electrons and protons are moved from the substrate on one side of the membrane by a coenzyme like plastoquinone to the other side of the membrane. Here there may be transfer of the electrons to an electron carrier like a cytochrome with accumulation of the protons within the membrane (Fig. 9.4). Proton translocation accompanying electron transport has been demonstrated with chloroplasts.[24] A pH gradient is formed during light-induced electron transport and the gradient is collapsed by detergents that presumably make the

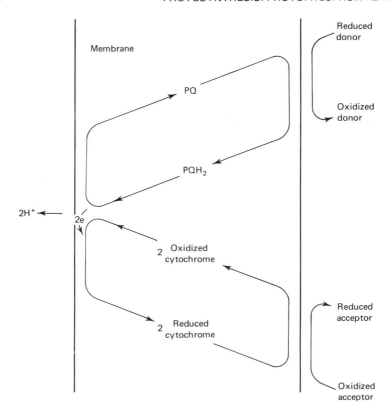

Figure 9.4   A model for proton transport across a membrane.

membrane freely permeable (Fig. 9.5). Generally, compounds that uncouple phosphorylation from electron transport collapse the proton gradient. The stoichiometry of protons translocated to electron transport is disputed, but values as low as 4 to 6 for the ratio of protons translocated per pair of electrons transported have been reported.[25]

Best evidence for the chemiosmotic hypothesis comes from the experiments of Jagendorf, in which exposure of chloroplasts to acid and then to base causes ATP synthesis in the dark. The chloroplasts are first acidified to pH 3.8 with succinic acid, which penetrates to the interior. The external pH is then adjusted up to 8 with base. Presumably, the succinate anions come out, leaving protons inside. ATP is certainly synthesized under these conditions and in a yield of 1 ATP per 5 chlorophylls.[26] If there is any relation in the number of phosphorylation sites to the number of electron carriers (1 PQ/10 chlorophylls or 1 cytochrome b/100 chlorophylls), the phosphorylation sites must be used repetitively in this experiment.

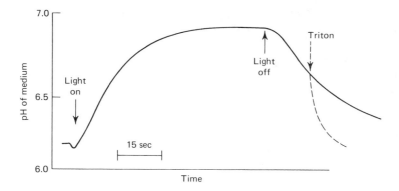

Figure 9.5   An experiment illustrating the ability of chloroplasts to accumulate protons in the light. The detergent Triton facilitates the
release of protons from the chloroplasts.

Correlative and more ambiguous evidence for the chemiosmotic hypothesis
comes from simultaneous measurement of ATP synthesis and change in pH.[27]
Such kinetic measurements show a lag in onset of ATP production at low light
intensities while the proton pump is functioning well at the low intensity. This
data might indicate that phosphorylation does not occur until a certain critical
pH gradient has built up.

Although proponents of the chemiosmotic mechanism hold that proton
gradients are the driving force for ATP synthesis, advocates of the "chemical
intermediate" school have maintained that proton gradients and ATP synthesis
are alternate expressions of the high-energy intermediate.

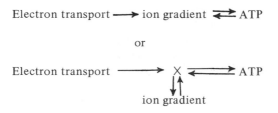

Chance and Avron have argued against the chemiosmotic theory with kinetic
data.[28],[29]

A serious difficulty with the chemiosmotic theory comes from the work of
McCarty, who treated chloroplasts with digitonin to produce small subparticles.[30],[31],[32] The subchloroplast particles carry out a cyclic photophosphorylation that is insensitive to uncoupling by ammonium ions. However,
ammonium ions abolish the proton transport activity of these preparations. Thus
pH gradient is not essential for photophosphorylation. In the presence of

potassium ions and the ionophorous antibiotic nigericin, the light-induced internal acidification of isolated chromatophores of *R. rubrum* is abolished, but photophosphorylation is unimpaired.[33] One can anticipate counterarguments to these observations.

Another approach to the problem of electron transport and photophosphorylation — regardless of the mechanism — is through the disassembly and reconstruction of a functional phosphorylating system. Avron was the first to accomplish an extraction-reconstitution experiment with photophosphorylation.[34] When chloroplasts were washed with EDTA, a chelator with especially high affinity for calcium ions, phosphorylation was eliminated and electron transport was increased as it would be by an uncoupler. When the EDTA in the supernatant fluid was complexed with excess calcium and the supernatant fluid added back to the washed chloroplasts, phosphorylation was restored and electron transport in the absence of ADP and inorganic phosphate was diminished, indicating that the system had been recoupled. Racker then showed that this protein, released from the chloroplasts by EDTA and called CF_1 (coupling factor), has a latent ATPase activity — presumably the reverse of its natural role in ATP synthesis.[35,36] Dithiothreitol activates the ATPase *in situ* or in the purified CF_1, while trypsin or heat treatment activates it *in situ* only. It is postulated that CF_1 is allotropic and, when membrane bound, functions as an Mg^{+2}-dependent ATP synthetase with high sensitivity to inhibitors. When solubilized, the same protein becomes a dithiothreitol activated, Ca^{++} dependent ATPase. There is evidence that CF_1 is a participant in the proton uptake process and there is a fascinating energy-linked conformational change of this protein.[37,38] Electron microscope studies show that CF_1 is a visible structure on the outer membrane surface of chloroplast particles.[39] There is evidence for a second protein factor — CF_2 — also required for photophosphorylation.[40] One looks forward to the time when these proteins will be identified more precisely than by the name "coupling factor."

In addition to these relatively orthodox views of photophosphorylation, one should recognize that Arnon, who discovered photophosphorylation, holds a very different view of the chloroplast mechanism.[40] Arnon maintains that Photosystem I is devoted exclusively to cyclic phosphorylation while Photosystem II involves two photo acts to transport electrons from water to NADP.

9.4   INHIBITORS AND PARTIAL REACTIONS OF PHOTOSYNTHESIS

The use of selective inhibitors and of partial reaction sequences has provided some of the best biochemical insights into the photosynthetic process.

The inhibitor DCMU, 3-(3,4-dichlorophenyl)-1,1-dimethylurea, was first recognized as a very potent herbicide that had little or no effect on heterotrophically grown plants. The site of DCMU inhibition was thus localized in the photosynthetic process and has been more precisely located near the oxygen evolving step. This location is assigned to DCMU, since it does not inhibit cyclic phosphorylation and since DCMU inhibition of NADP reduction can be reversed by using reduced indophenol dye as an electron donor in place of water. Since DCMU causes an immediate rise in Photosystem II fluorescence, it is postulated that DCMU causes the traps of Photosystem II to fill rapidly in the light. Since the inhibitor prevents the traps from being drained by chemical oxidation, the absorbed energy must be lost as fluorescence. Izawa and Good have done a careful study of the amount of DCMU needed to cause inhibition and concluded that there is one DCMU-sensitive site for every 2,500 chlorophylls, which is in remarkable agreement with the value of the photosynthetic unit of Emerson and Arnold.[41] By implication, DCMU reacts with a component that is present in exceedingly low concentration.

Katoh and San Pietro have found a second and distinct inhibitable site in the oxygen evolving region of the photosynthetic chain. In this case the inhibition is the result of heating algal chloroplasts and it apparently works by denaturing a singularly heat-labile component.[42] In these experiments, chloroplasts from *Euglena* were heated for five minutes at $40°C$ and this treatment abolished Hill reaction activity. NADP photoreduction could be restored by adding indophenol dye and either ascorbate or cysteine to keep the dye reduced. In fact, activity could be restored by either ascorbate or cysteine without the dye — this in contrast to the case of DCMU inhibition in which the dye is essential in mediating the transfer of electrons from ascorbate into the chloroplast chain. The activity of the heated chloroplasts in taking electrons from ascorbate or cysteine to NADP was blocked by DCMU and, of course, this DCMU inhibition could be reversed by adding indophenol dye. The interpretation of these results is diagrammed in Figure 9.6. These data do not show which side of Photosystem II the electrons enter, but they provide clear evidence for a new site of inhibition and a new site of electron donation. Yamashita and Butler have developed a similar set of experiments for spinach chloroplasts that have been inhibited by washing with high concentrations of Tris buffer.[43,44] In these experiments paraphenylene diamine was needed to mediate the flow of electrons into the chloroplast electron transport chain at a site which is still sensitive to DCMU inhibition (Fig. 9.7). Subsequently, Yamashita and Butler found that heat treatment and UV irradiation block spinach chloroplast electron transport at approximately the same site as the Tris washing.[45]

Photosystem II is selectively inhibited by a great variety of manipulations. Among the more recent are digestion with trypsin[46] or with lipase;[47] treatment with hydroxylamine,[48] carbonyl cyanide phenylhydrazone,[49,50] hydroxyquinoline-N-oxime;[51] and by mutations.[52] In some of the above situations, inhibition can be reversed by using artificial electron donors to Photosystem II. Manganous ion,[53] hydrogen peroxide,[54] ascorbic acid,[55] hydrazine,[56] hydrox-

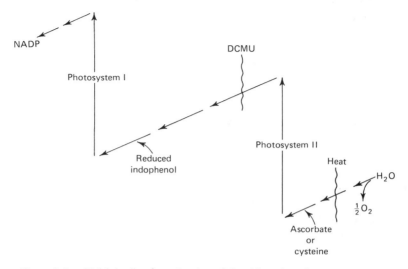

Figure 9.6   Multiple sites for reduction of the chloroplast electron-transport chain by artificial electron donors.

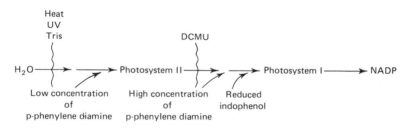

Figure 9.7   Multiple sites for the entrance of electrons from artificial donors into the chloroplast electron-transport chain.

ylamine,[57] and diphenyl carbazide[58] will donate electrons to Photosystem II. It seems likely that by checking the various kinds of electron donors against the various inhibitors of Photosystem II, empirical evidence will accumulate for multiple steps in this part of the electron transport chain.[59]

Specific inhibitors for Photosystem I are less abundant. Berzborn obtained an antibody that inhibited ferredoxin-NADP oxidoreductase and another that appears to inhibit the primary reductant of Photosystem I.[60] The fact that the ferredoxin-NADP oxidoreductase antibody did not agglutinate the chloroplasts led to the inference that the antigen must be in a crevice in the lamellar surface, which is sufficiently deep to prevent the two combining sites on the antibody from cross-linking two antigens. As noted earlier, the plastocyanin antibody has been most instructive in demonstrating the role of plastocyanin in cyclic phosphorylation and in indicating the buried position of plastocyanin in subchloroplast membrane vesicles. Both of these antibody preparations indicate the special value of this technique in studying the molecular topography of chloroplast structure. Large polycations such as histone and polylysine are specific inhibitors of Photosystem I, and their use indicates that while low redox potential electron acceptors are reduced exclusively by Photosystem I, acceptors with a high redox potential such as indophenol dyes and ferricyanide are reduced by both photosystems.[61]

9.5   SUBUNITS OF CHLOROPLAST STRUCTURE AND FUNCTION

Fractionation of the photosynthetic apparatus has led to evidence concerning discrete functional units within the chloroplast membranes. Boardman observed that incubation of chloroplasts with digitonin, followed by differential centrifugation, gave two types of subparticles:

> 1. A small particle enriched for chlorophyll a that catalyzed the photoreduction of NADP with reduced indophenol dye as donor.[62] Compositional analysis showed that this particle has most of the cytochrome f and b_6 and most of the P700, but, oddly enough, it contains most of the PQ as well.
>
> 2. A large particle enriched for chloroplyll b that will photoreduce indophenol presumably as a minimal expression of Photosystem II. The large particle shows the Photosystem II fluorescence phenomena and is enriched for manganese and cytochrome b_{559}.

This split of the chloroplast into two functionally distinct subunits is not absolutely clean, but the ratio of 1 P700/200 chlorophylls in the small Photosystem I particle and 1 P700/700 chlorophylls in the large Photosystem II particles is an indication that the small particles are nearly pure Photosystem I and that the large particles are 70% Photosystem II and 30% Photosystem I. Mechanical disruption, as well as a large variety of detergents, have been used to split apart the two photosystems.[63] Vernon tried this technique on blue-green algae using Triton X-100 as a dissociating detergent on particles that had been freed of phycocyanin by washing.[64] The disrupted material gave two bands of particles on density gradient centrifugation. Both bands contained chlorophyll a, but only one had P700. While absorption spectra showed modest differences in the carotenoid content, fluorescence spectra showed interesting differences in the chlorophyll a in the two particles. These spectra are shown in Figure 9.8. The chlorophyll a in Photosystem I fluoresces at liquid nitrogen temperature at 731 nm while the chlorophyll in Photosystem II emits light at 678 to 685 nm, indicating that these two forms of chlorophyll a are at least in very different environments. This work provides a valuable criterion for the recognition of two different chlorophyll complexes.

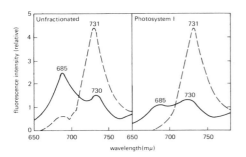

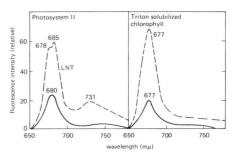

Figure 9.8   Fluorescence emission spectra of photosynthetic membranes showing the preponderance of the 731 nm fluorescing form of chlorophyll a in Photosystem I. [From T. Ogawa et al., *Biochim. Biophys. Acta*, **172**, 216 (1969).]

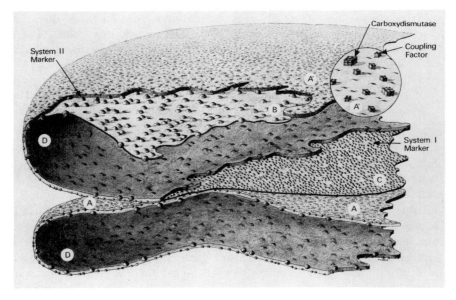

Figure 9.9   A schematic representation of two chloroplast thylakoids showing a binary membrane structure. Face B revealed by freeze-etching contains the large 175 Å particles. The 10,000 *g* digitonin fraction is enriched in these particles and in Photosystem II activity. Face C contains the 100 Å particles seen by freeze-etching. The 144,000 g digitonin fraction is characterized by these particles and by having high Photosystem I activity. [From Arntzen et al., *J. Cell. Biol.*, **43**, 16 (1969).]

In an interesting extension of this work, Briantais has described the splitting of higher plant chloroplasts with Triton into two photosystem substructures and, on removal of the detergent, he found that the subparticles would reaggregate.[65] Arntzen et al. reassembled Photosystem I and Photosystem II subparticles with a restoration of electron transport activity that involved the combined actions of both photosystems.[66] The relation of the functionally distinct subfractions to distinct morphological structures within the chloroplast promises to provide a detailed bridge between molecular composition and structures made visible by electron microscopy.[63,67] Figure 9.9 is a diagram summarizing electron microscopic evidence on the disposition of various recognizable components on the chloroplast membrane.

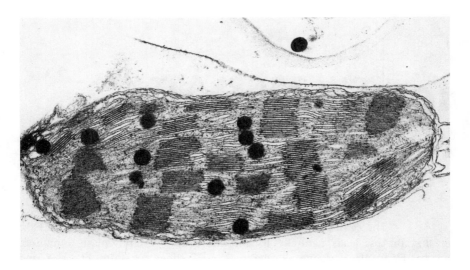

Figure 9.10   Chloroplast of a mesophyll cell of corn (92,000 ×). (Courtesy of Drs. J. Hall and R. Barr.)

10 ‖ CHLOROPLAST DEVELOPMENT

The earlier discussion of chlorophyll biogenesis and the role of light in that process leads naturally to a consideration of chloroplast formation. Leaves from plants grown in darkness are yellow to white in color and microscopic examination reveals proplastids of very different structure from the mature chloroplasts. When brought into the light, the pigment formation, morphological changes, and acquisition of photosynthetic competence involve a number of biochemical processes under precise control.[1]

10.1 PHOTOCONVERSION OF PROPLASTIDS

Morphological development starts with a proplastid which contains a large paracrystalline body. A brief flash of light (which also converts protochlorophyllide to chlorophyllide) will cause a morphological change called "tube transformation" that dissociates the paracrystalline material into loosely packed vesicles in a few seconds.[2] This change in structure appears to be initiated by a single photochemical step. Perhaps the converted chlorophyllide holochrome directs the assembly of the first lamellar membrane elements. After five hours of illumination, these vesicles disperse and then fuse to make the primary lamellae. Later, additional discs form along the primary lamellae to make the mature grana stacks.[3]

10.2   ENZYME CHANGES ASSOCIATED WITH GREENING

Biochemical development has been detailed first in the protochlorophyllide conversion — the immediate photoconversion gives only 0.3% to 1% of the final chlorophyll. However, even a brief flash of light will stimulate a tenfold increase in protochlorophyllide formation in the subsequent dark period. This flash can be replaced by feeding leaves ALA (delta amino levulinic acid) in the dark, indicating that all the necessary enzymes are present to convert ALA to protochlorophyllide. The flash must cause an increase in ALA synthesis in order to accelerate the protochlorophyllide accumulation.

Sustained illumination will cause a transient rise in the succinyl CoA synthetase activity reaching a maximum at 6 hours illumination and returning to the dark level at 12 hours illumination.[4] This enzyme transient may facilitate a temporary mobilization of succinate for chlorophyll formation. There is also an increase in the ALA dehydrase in greening bean leaves in continuous light.[5] Since this activity is not induced by a flash and is not reversed by far red light, it apparently is not under phytochrome control.

The flash-induced increase in ALA synthetase is the result of *de novo* synthesis of the enzyme as indicated by the fact that protein synthesis inhibitors prevent the light flash from inducing accumulation of protochlorophyllide. Bogorad has followed these activities by spectral assay of the 650 nm protochlorophyllide and 675 nm chlorophyll peaks in intact leaves before and after illumination. His results are illustrated in Figure 10.1. Thus, puromycin and chloramphenicol (inhibitors of protein synthesis) block protochlorophyllide synthesis. Similar effects are obtained with actinomycin D, which blocks DNA-dependent RNA synthesis. Light may be derepressing the ALA synthetase gene (and the messenger may be translated on a chloroplast ribosome, since the synthesis of this protein is chloramphenicol–sensitive). In maize, ALA formation responds very promptly to light, and other enzymes associated with the photosynthetic process respond with varying degrees of promptness of induction on illumination.[6] This experiment is similar to a test for coordinate derepression of a bacterial operon. Light causes an immediate appearance of ribulose diphosphate carboxylase, but the kinase and isomerase activities do not appear until hours later. The appearance of all three enzymes can be blocked by pretreatment with chloramphenicol. Therefore, enzyme appearance seems to be the result of new protein synthesis.

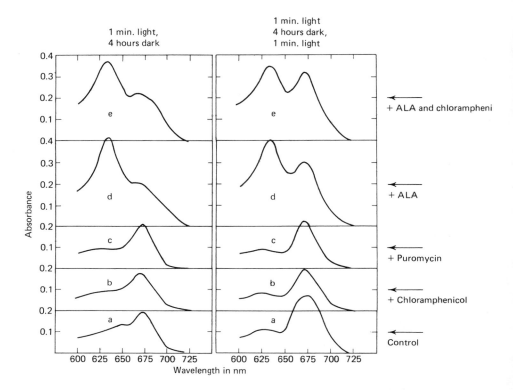

Figure 10.1   Absorption spectra of dark-grown kidney bean leaves taken
after exposure to light and after treatment with inhibitors of
protein synthesis. The accumulation of protochlorophyll is
prevented by the inhibitors, and this inhibition is overcome by
ALA. [Redrawn from Gassman, M. and Bogorad, L., *Plant
Physiol.*, **42**, 781 (1967).]

McMahon and Bogorad tested the role of photosynthesis in the appearance of
the "early light" and the "late light" enzymes.[7] Pretreatment of the leaves with
CMU to inhibit photosynthesis did not inhibit the synthesis of ribulose
diphosphate carboxylase, but it did block the appearance of the ribose 5
phosphate isomerase. Total soluble protein in the leaf was unaffected by the
CMU treatment, indicating that at this stage of development the amount of
protein synthesis dependent on illumination is not a large fraction of the total
soluble protein in the cell. Feeding glucose to the leaves did not reverse CMU
inhibition of isomerase appearance, which may indicate that this inhibition is
not simply a result of carbon starvation. One is left with the suggestion that a
product of photosynthesis is a specific derepressor of the isomerase. A number
of protein synthesis inhibitors will block the rise in ribulose diphosphate
carboxylase and chlorophyll but not the rise in ribulose phosphate kinase in

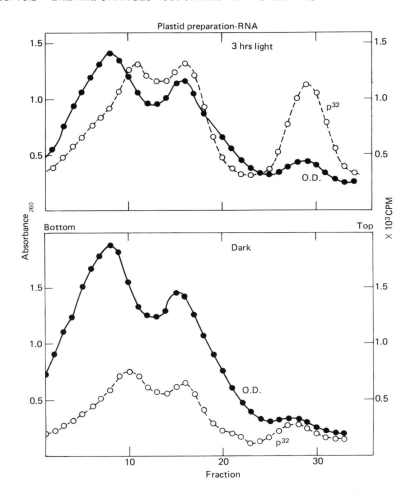

Figure 10.2   Density-gradient centrifugation analysis of chloroplast RNA. After three-hours illumination of etiolated leaves, there is a greatly increased incorporation of P^{32} into several size classes of RNA. (From Bogorad, L., in *Harvesting the Sun*, A. San Pietro, F. A. Greer, and T. J. Army, eds., Academic Press, N.Y., 1967, p. 191.)

greening barley — suggesting a separate translation mechanism or the activation of preformed kinase in this species.[8] Although both cycloheximide and chloramphenicol inhibited ribulose diphosphate carboxylase appearance[9] (when these inhibitors should distinguish between cytoplasmic and chloroplast ribosomes), this may be because of the synthesis of the two different subunits of the carboxylase on two different kinds of ribosomes.[10,11]

In any case, the actinomycin inhibition of greening indicated a dependence on RNA synthesis and led to other evidence for enzyme induction at the nucleic acid level. Etiolated maize was fed P^{32} or C^{14} uridine and then either kept in darkness or exposed to light. After being illuminated with isotope, the RNA was extracted and run through a density gradient centrifugation. Plastid RNA shows up to sixfold increase in specific activity – in all size classes of RNA – as a result of illumination. This is in comparison to very slight increases in cytoplasmic RNA in response to light. The effect of light on plastid RNA is inductive in that a 30-minute illumination plus 90 minutes in darkness gives the same effect as 120 minutes of light (Fig. 10.2).

Since all size classes of RNA are synthesized, Bogorad sought an increase in DNA-dependent RNA polymerase and found that while whole-leaf homogenate polymerase activity showed no difference for dark or light treated leaves, washed plastids showed a definite increase in polymerase activity. This increase in RNA polymerase activity could be inhibited 80% by chloramphenicol, suggesting that plastid ribosomes translate the messenger for plastid RNA polymerase. These observations are summarized in Table 10.1.

TABLE 10.1   TIME OF INITIAL INCREASE IN CHLOROPLAST
                 COMPONENTS

Seconds	*Minutes*	*Hours*
Protochlorophyllide to chlorophyllide	RNA production RNA polymerase	ribulose phosphate kinase
Tube transformations	ALA production ribulose diphosphate carboxylase	ribulose phosphate isomerase

10.3   PHOTOCONTROL OF CHLOROPLAST DEVELOPMENT

The nature of the photochemical act is not clear beyond the protochlorophyllide holochrome conversion that involves 650 nm light absorption. Possibly this conversion, with the attendant changes in the conformation of the holochrome protein, causes the rapid morphological reorganization that is recognized as tube transformation. Further changes in enzyme content may be phytochrome mediated, but not all of these changes show far red reversal of the initiating red light induction. It was noted earlier that NADP-linked triose phosphate dehydrogenase is under phytochrome control. Phytochrome activation by brief exposure of etiolated pea seedlings to red light will cause a marked increase in the ribosomal RNA content of the etioplast. This activation, in which the red light effect is nullified by prompt subsequent exposure to far red light, can raise the level of ribosomal RNA found in the chloroplast precursor structure to that found in a fully green, mature chloroplast. Klein has examined the phytochrome control of protein synthesis in primary leaf pairs of bean.[12] Phytochrome activation by red light is followed by a general increase in total protein, and this includes enzymes of chloroplasts, mitochondria, and cytoplasm. This is in contrast, but not necessarily in conflict, to Bogorad's finding that new RNA synthesis was confined to the plastid. Klein found that phytochrome-induced protein synthesis was not blocked by fluorouridine deoxyriboside, which does inhibit DNA synthesis. Thus, the phytochrome-induced increase in protein is not the result of cell division.

The appearance of photosynthetic electron transport and phosphorylation activities with greening is an interesting sequence. Although chlorophyll b appears ten minutes after the beginning of illumination in greening pea seedlings, the appearance of Hill reaction activity lags by several hours and is better correlated with the formation of grana.[13,14] Thus the assembly of the photosystems is not a single-step, concerted process. If bean plants are exposed to red light, chlorophyll a is synthesized continuously, but chlorophyll b does not appear until the fifth day of such illumination. Oxygen evolution and photophosphorylation activities precede the appearance of both chlorophyll b and the formation of grana, which do not develop at all until the plants are exposed to white light.[15] These observations suggest that several different photocontrols of development must be kept in balance. In the greening bean chloroplast cyclic phosphorylation can precede the appearance of Hill reaction activity,[16] but even this process, which is centered on Photosystem I, must develop piecemeal. The CF_1 coupling factor that seems an intergral part of the chloroplast phosphorylation mechanism is present in the etioplast of plants that have never seen light and the phosphorylation activity induced by acid-base transitions can develop in the dark after a brief exposure to light.[17,18,19]

The question is surely an open one, but phytochrome action may derepress some of the "early light enzymes" which, after several hours of operation, cause the buildup of products of photosynthesis, which in turn derepress the "late light enzymes." When a complete list of "early light," "late light," and "dark constitutive" enzymes is available, we may be better able to sort out the control mechanisms.

10.4   METABOLIC CONTROL AND MEMBRANE ASSEMBLY

Among the algae, there is evidence for metabolic control related to carbon metabolism. Some strains of *Chlorella* that normally remain green during heterotrophic growth in the dark will become bleached during growth on very high concentrations of glucose, suggesting a possible catabolite repression of photosynthetic machinery. *Euglena* normally bleaches in darkness and regreens in light. The inclusion of a carbon source in the usual autotrophic medium used for *Euglena* inhibited regreening.[20] Kirk found that the carbon source used for heterotrophic dark growth of *Euglena* was best for inhibiting regreening, suggesting that adaptation to a specific carbon source in the dark makes that compound a better repressor in the light.[21] Kirk later showed that the repressor effect could be reversed by ammonium sulfate in the medium, presumably by draining off the excess carbohydrate catabolite and converting it to protein.[22] These experiments have been elaborated on in order to show that a high carbon-to-nitrogen ratio in the medium is responsible for inhibition of regreening.[23]

There are detailed studies of regreening of *Euglena* in the shift to autotrophic growth.[24] Inclusion of DCMU to completely inhibit carbon dioxide fixation allows 70% of the normal level of chlorophyll synthesis, and the chloroplasts look normal in the electron micrographs. Thus chloroplast development does not require photosynthesis, and one role of light is to activate the utilization of cellular reserves for assembly of the photosynthetic structure. Dark-grown *Euglena* does contain RNA that is complementary to chloroplast DNA and on illumination there is a rapid methylation of this preformed RNA.[25,26] It seems likely that only a fraction of the chloroplast genome is under photocontrol and some chloroplast genes are transcribed and perhaps translated in the proplastid.

The photochemical activation step and even the photoreduction of protochlorophyllide are not present in many algae and gymnosperms that develop apparently complete chloroplasts in the dark.

In higher plants, the plastid protein and lipid increase threefold during greening with little change in the relative concentrations (50–65% lipid and 20–25% protein) of these two components. You will recall that chloroplasts contain enzymes for the synthesis of chloroplast lipids. The appearance of specific lipid classes, especially in relation to structural development of the chloroplast membranes, might prove very instructive in understanding the forces of membrane assembly.[27]

Hoober et al. have presented interesting data on the greening process in *Chlamydomonas reinhardi.*[28] Chloramphenicol, which is presumed to selectively inhibit the function of chloroplast ribosomes, allows the greening and membrane formation but inhibits the appearance of electron transport activity. The chlorophyll containing discs are not fused to give the usual grana stacks in the presence of chloramphenicol. Cycloheximide, presumably a selective inhibitor of cytoplasmic ribosomes, inhibits greening and membrane formation by about 50%, but there is disc fusion and indophenol dye reduction by the Hill reaction with these chloroplasts. Each of these drugs inhibits leucine incorporation into the cells by about 50% and their effects are additive. Thus both cytoplasmic and chloroplast ribosomes contribute proteins to the synthesis of chloroplasts. Since there is no change in the buoyant density of the photosynthetic lamellae during greening of *Chlamydomonas*, the assembly may be random insertion of new materials within the pre-existing membrane rather than stepwise assembly of different membrane layers.[29] In contrast, *Euglena* chloroplast membranes may form in a discontinuous fashion with an early preponderance of lipid.[30]

Of great interest is the observation that carefully prepared chloroplasts can divide after isolation from the leaf cell.[31] Although this *in vitro* division need not involve massive synthesis of chloroplast constituents, one could hope to identify the components involved in the terminal stages of organelle assembly.

10.5   CHLOROPLAST AUTONOMY

The greening and chloroplast maturation phenomena mentioned earlier suggest a certain temporal autonomy of some genes involved in chloroplast synthesis. A fraction of the genes directing the synthesis of chloroplast components are extranuclear and are located within the chloroplast.[32] Some of the best evidence for genetic autonomy of chloroplasts comes from old observations on variegated plants:

> Flowers on green branches make seed which gives rise to green plants;
> Flowers on variegated branches make seed which gives rise to variegated plants;
> Flowers on colorless branches make seed which gives rise to colorless plants;

and in all cases the source of pollen makes no difference. Thus the inheritance is maternal and controlled by genes in the cytoplasm of the egg.

More recent evidence of chloroplast autonomy is seen in the bleaching of *Euglena* by streptomycin. Exposure to streptomycin during autotrophic growth is fatal, but exposure during heterotrophic growth gives cells that have permanently lost the ability to make chloroplasts. This is the result of selective inhibition of chloroplast reproduction. Since there is microscopic observation of both chloroplasts and proplastids dividing, these organelles clearly replicate independently of cell division.

10.6   CHLOROPLAST DNA

Chloroplasts are now conceded to contain DNA (0.5% of the dry weight) after much controversy about contamination. The main evidence that this DNA is really in the chloroplast comes from satellite banding of DNA in ultra-centrifuge analyses and from tritiated thymidine incorporation into chloroplasts as measured by autoradiography. Early attempts at DNA analysis by density gradient fractionation indicated a DNA band in the whole cell DNA that was not seen in the nuclear DNA, and this band was enriched by isolating DNA from the chloroplast fraction. When chloroplasts are prepared by differential centrifugation, the DNA is badly contaminated with nuclear DNA. Isolation of chloroplasts by sucrose density gradient procedures minimizes the nuclear contamination. Figure 10.3 shows data from a typical experiment. Buoyant density measurements reveal a difference between chloroplast DNA and

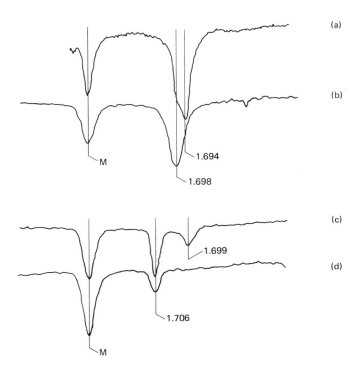

Figure 10.3   Photoelectric scans of DNA after density-gradient centrifugation in cesium chloride. The centrifugation time was 20 hr at
44,000 rpm at 20°C with *Micrococcus lysodeikticus* DNA of
density 1.731 as marker. (a) DNA from crude chloroplast
fraction; (b) DNA from fraction (a) after DNase treatment;
(c) DNA from crude mitochondrial fraction; (d) DNA from
fraction (c) after DNase treatment (unpublished results).
[From Tewari, K. K., *Ann. Rev. Plant Physiol.*, **22**, 141
(1971).]

mitochondrial DNA from the same cell.[33] Representative data are found in Table 10.2. Tobacco chloroplast DNA is distinguished from nuclear DNA in that it is devoid of 5 methylcytosine.[34] Carefully prepared chloroplast DNA from broad bean appears to have the same buoyant density as nuclear DNA.[35] In this species, the chloroplast DNA can be recognized by its better renaturation behavior, suggesting more sequence repetitions. On a weight basis, the nucleus of the broad bean leaf cell contains 500 to 1,000 times as much DNA as is found in

TABLE 10.2   BUOYANT DENSITY

	Mitochondrial DNA	Chloroplast DNA
Pea[2]	1.705	1.694–1.695
Tobacco[3]	1.706	1.697

the chloroplast. Chloroplast DNA appears conventional in $A/T = G/C = 1$, melting behavior, and double strandedness. Chloroplast DNA replicates by the usual semi-conservative mechanism.[36]

An interesting question is the amount of DNA per chloroplast and the extent of its function. Most estimates for higher plants and algae run from 2 to 10×10^{-15} grams DNA per chloroplast, which is close to the 7×10^{-15} grams DNA found in a bacterial cell. This seems to be a lot of informational capacity when compared to the mitochondrion. There is convergent data from the molecular weight estimates of 4 to 11×10^{7} for chloroplast DNA and from autoradiography that there are approximately 20 copies of one DNA molecule per chloroplast.[37,38,39] If these 20 DNA molecules are identical, the total coding capacity shrinks to fewer different species of proteins coded in the chloroplast DNA. This number would shrink further if there are redundant nucleotide sequences within the DNA and should certainly be adjusted downward by a few percent of the nucleotides that are used to code chloroplast transfer RNA and chloroplast ribosomal RNA.

There is evidence for at least two distinct species of DNA in some chloroplasts, suggesting two different linkage groups or chromosomes in this organelle.[40,41,42] A large, circular molecule of DNA is found in *Euglena* chloroplasts.[43]

In addition to DNA, the chloroplast contains the machinery for gene replication and expression. There is a DNA polymerase activity, requisite for gene duplication, which can be detected in gradient washed chloroplasts.[44,45,46] Both the enzyme and the DNA template are tightly bound to the chloroplast, but a radioactive product can be isolated. Product formation is dependent on all four deoxyribonucleoside triphosphates being present as substrates and is inhibited by DNAase and by actinomycin D. The C^{14} labelled product bands in a $CsCl_2$ gradient like chloroplast DNA and hybridizes with chloroplast DNA (45%) but hardly at all with nuclear DNA (7%), indicating that this chloroplast polymerase is directed by a chloroplast DNA template.

Gene expression in the chloroplast is presumably accomplished by DNA-dependent RNA polymerase. This enzyme has been observed in chloroplasts and partially characterized.[47,48,49] Although *Chlamydomonas* chloroplast DNA transcription is sensitive to selective inhibition by rifampicin, the chloroplast polymerases of higher plants are not inhibited by this antibiotic and are thus distinguished from the analogous bacterial enzymes.[50,51]

10.7   CHLOROPLAST RIBOSOMES

Chloroplasts are seen to contain ribosomes in electron micrographs and when these ribosomes are isolated, they, like mitochondrial ribosomes and bacterial ribosomes, are 70S in contrast to the 80S ribosomes found in the cytoplasm of higher plants and animals. Polyribosomes are found bound to the chloroplast lamellae and are quite active in amino acid incorporation.[52] When ribosomal RNA is isolated from the chloroplast ribosomes, it hybridizes readily with both chloroplast DNA and nuclear DNA.[53,54] In contrast, cytoplasmic ribosomes give a ribosomal RNA that hybridizes with nuclear but not chloroplast DNA. Thus the chloroplast DNA may code for only a small number of highly specific ribosomes within the chloroplast. Some of the chloroplast ribosomal RNA sequences may occur in the DNA of both the nucleus and the chloroplast through redundancy and/or something like translocation of DNA among chromosomes. Two types of mutations alter the chloroplast ribosomes of *Chlamydomonas*. Several non-Mendelian, uniparentally inherited mutations that alter the response of this alga to antibiotics are correlated with alterations in the appearance of the chloroplast ribosomes.[55] A mutation showing Mendelian segregation and, therefore, presumed to be in a nuclear gene causes accumulation of the large subunits of chloroplast ribosomes, perhaps because of deficiency in the assembly of the small ribosome subunit.[56] This latter mutation suggests but does not demand participation of a nuclear gene in chloroplast ribosome assembly, since deficiency in a nuclear gene product might create a metabolic imbalance that simply interferes with the expression of chloroplast genes.

Careful characterization of ribosomes from chloroplasts has yielded the structural contrasts that are summarized in Table 10.3[57,58,59] As with the

TABLE 10.3   CONTRASTING ELEMENTS IN
              RIBOSOME STRUCTURE

80S ribosome in animal cytoplasm containing:
         29S RNA with associated 5.8S and 5S RNA
         18S RNA

80S ribosome in plant cytoplasm containing:
         25S RNA with associated 5.8S and 5S RNA
         16S RNA

70S ribosome in bacteria containing:
         23S RNA with associated 5S RNA
         16S RNA

70S ribosome in chloroplasts containing:
         23S RNA with associated 5S RNA
         16S RNA

other types of ribosomes, the chloroplast ribosome is dissociated by removal of magnesium ion. Chloroplast rRNA differs from cytoplasmic rRNA in base composition, and the 23S rRNA is uniquely labile during isolation.[60] Isolation of chloroplast DNA and chloroplast 16S and 23S ribosomal RNA from pea leaves has permitted some interesting hybridization studies.[61] About 4% of the chloroplast DNA is devoted to ribosomal RNA coding and there appear to be two cistrons each for the 16S and 23S RNA. In *Chlamydomonas reinhardi*, the chloroplast ribosomal 16S and 23S RNA are encoded in tandem and there are multiple cistrons here as well.[62]

Like ribosomes from other sources, the chloroplast ribosomes are only 44% RNA and the rest is protein. Lyttleton isolated ribosomes from spinach chloroplasts, dissociated them in urea, and separated about 15 protein bands on gel electrophoresis (c.f. 30 bands from *E. coli* ribosomes).[63,64] The ribosomal protein pattern differs markedly among ribosomes from cytoplasm, chloroplasts, and mitochondria.[65] The pattern of chloroplast ribosome protein varies with the taxonomic position of various types of plants tested, but ribosomal proteins do not vary among different organs of the same plant. This argues against a control of differentiation because of the different ribosome structures. Again, in comparing different species, it appears that the cytoplasmic ribosome patterns are more similar than the patterns for chloroplast ribosome proteins.[66]

The isolation of a specific messenger RNA is a difficult task, but Bogorad has shown incorporation of P^{32} into a variety of RNA species from chloroplasts that had been isolated from leaves that had been given a brief illumination to initiate development of photosynthetic competence.[67] Light-initiated synthesis of mRNA in greening *Euglena* chloroplasts is well characterized.[68,69]

tRNA has been isolated from chloroplasts and appears to be distinct from cytoplasmic tRNA.[70] Since a characteristic of bacterial protein synthesis is the use of formyl methionine as the initial amino acid in the polypeptide sequence, it is not surprising that the bacterial type ribosomes of chloroplasts should use this initiator also. N-formyl methionyl tRNA has been found in wheat and bean chloroplasts.[71,72] Some of the leaf tRNA shows preferential hybridization with chloroplast DNA, indicating that chloroplast genes code chloroplast tRNA.[73] In greening *Euglena*, the light-induced appearance of chloroplast tRNA has been observed.[74] Finally, one might expect a set of amino acid activating enzymes in the chloroplast and there is increasing data on these constituents. Several chloroplast tRNAs cannot be charged by cytoplasmic aminoacyl tRNA synthetases but only by enzymes isolated from chloroplasts.[75,76] A bleached mutant of *Euglena* has been found that contains an aminoacyl tRNA synthetase that is specific for a chloroplast tRNA, yet this mutant has no detectable chloroplast DNA.[77] This implies that this amino acid activating enzyme is a nuclear gene product.

Thus the chloroplast appears to have its own distinctive DNA, DNA polymerase, RNA polymerase, ribosomes, tRNA, and amino acid acitivating enzymes. One returns to the question of the degree of autonomy of the chloroplasts. There is evidence that the chloroplast DNA codes rRNA and tRNA. It would be reasonable to have all of the ribosomal proteins of the chloroplast encoded in the chloroplast DNA, but there is no proof of this as yet.

Chloroplasts contain most, if not all, of the enzymes to make themselves — systems for the synthesis of fatty acids and galactolipids, carotenoids, chlorophyll, etc., as well as the protein within the chloroplast. How many of these proteins are encoded in the chloroplast genome? Wherever a characterized mutant is available, the gene locus can be tested, since nuclear genes segregate on a Mendelian basis but chloroplast genes do not. A number of mutations in chlorophyll and carotenoid synthesis have been identified with nuclear genes, as have the mutants defective in photosynthetic electron transport in *Chlamydomonas*. The mutations in variegated plant tissue are poorly characterized and might well represent structural defects in chloroplast assembly or backbone protein that prevents proper placement of pigments and causes them to be photooxidized as rapidly as they are formed.

As with mitochondrial ribosomes, the chloroplast ribosomes show an inhibitor sensitivity different from that of cytoplasmic ribosomes. Aminotriazole, a frequent herbicide constituent, seems to selectively inhibit the formation of chloroplast ribosomes.[78] Ingle has done extensive studies of ribosome function in higher plant leaves (radish cotyledon).[79] Chloramphenicol, streptomycin, and cycloheximide inhibited the chloroplast ribosome appearance in greening cotyledons more than they inhibited the appearance of cytoplasmic ribosomes. Inhibition of chloroplast ribosome synthesis by cycloheximide suggests that cytoplasmic ribosomes contribute to this synthesis.[80] This may be related to the cycloheximide inhibition of chloroplast DNA synthesis.[81] Chloramphenicol inhibits the synthesis of nitrite reductase, a chloroplast enzyme, but not nitrate reductase, a cytoplasmic enzyme.[82] Chloramphenicol inhibits synthesis of ribulose diphosphate carboxylase and NADP-linked glyceraldehyde phosphate dehydrogenase. Therefore, all of these chloroplast enzymes may be assembled on chloroplast ribosomes, but this does not demand a chloroplast origin for these mRNAs. Alternately, the chloramphenicol inhibition may mean that assembly of these enzymes on cytoplasmic ribosomes requires some product or signal from functioning chloroplast ribosomes.

Smillie and Linnane have each demonstrated a similar situation in *Euglena* — chloroplast ribosomes sensitive to chloramphenicol and cytoplasmic ribosomes sensitive to cycloheximide.[83,84,85] Chloramphenicol does not interfere with chloroplast DNA synthesis in *Euglena*. This again suggests that chloroplast DNA synthesis may be dependent on cytoplasmic ribosome products.[86] There is selective binding of C^{14} chloramphenicol to the chloroplast ribosome.[87] Finally, the chloroplast ribosome has been shown to be specifically inhibited by D-threo-chloramphenicol, the only one of the four stereoisomers active against bacterial ribosomes.[88]

A similar selectivity of inhibition is shown by the antibiotics spectinomycin, lincomycin, and erythromycin, which inhibit amino acid incorporation by chloroplast but not by cytoplasmic ribosomes.[89] These inhibitors are valuable probes at the sites of synthesis of chloroplast proteins.[90]

The chloroplast shows many similarities to bacteria in terms of sensitivity to specific antibiotic inhibitors, ribosome structure, and initiation of polypeptide formation. There is an inclination to assume from these similarities that the chloroplast evolved from a symbiotic bacterium living within the plant cell. This assumption will need more evidence before one can rule out a parallel or repetition of evolution of bacteria-like catalysts somehow well fitted to the chloroplast's needs. Pigott and Carr have found significant homology between the ribosomal RNA from several species of blue-green algae and the chloroplast DNA of Euglena.[91] This shows genetic homology between procaryotes and the chloroplast genes of a eucaryote. One can expect that more molecular data of this kind will clarify the evolution of photosynthetic structures.

GENERAL REFERENCES _____

Kirk, J. T. O. "Biochemical Aspects of Chloroplast Development," *Ann. Rev. of Plant Physiol.*, **21**, 11 (1970).

Kirk, J. T. O., and R. A. E. Tilney-Bassett. *The Plastids, their Chemistry, Structure, Growth and Inheritance*. San Francisco: W. H. Freeman and Co., 1967.

Tewari, K. K. "Genetic Autonomy and Extranuclear Organelles," *Ann. Rev. of Plant physiol.*, **22**, 141 (1971).

11 PHOTOCHEMICAL AND HORMONAL CONTROLS

Both light as measured in terms quite distinct from photosynthesis and hormonal chemicals can produce profound physiological responses in plants. Biochemical investigations have helped to clarify the nature of the substances that initiate these responses. Phytochrome is the pigment that measures day length and prompts the plant to respond properly to the physiological demands of the season.[1] The molecular identity of phytochrome is well in hand. A large number of plant hormones and of synthetic growth regulators are known. The responses to phytochrome and to the chemical regulators are known in physiological terms, e.g., the plant flowers, and sometimes in rather coarse biochemical terms, e.g., an enzyme is synthesized or a membrane charge is altered. However, the molecular details that intervene between perception of the stimulus and final response are generally obscure. What follows is a very incomplete catalog of biochemical information about the regulation of plant growth and development. Hopefully, the missing biochemical terms will begin to fall in place around the scant facts listed here.

11.1   PHYTOCHROME

Phytochrome has been mentioned as a receptor pigment in the photoregulation of chloroplast development and in the appearance of photosynthetic enzymes. Phytochrome has been under long and honorable investigation at the USDA Beltsville laboratories where Garner and Allard first made a clear case for photoperiodic control of flowering. Borthwick and his colleagues found an enormous number of physiological expressions of photoperiod control – stem elongation, leaf expansion, plumular hook opening, spore and seed germination, regulation of flowering in many species, and dormancy control. Hendricks examined the quality of light effective in photocontrol and his measurements of the action spectra of the various plant responses revealed that the same promoting and inhibiting wavelengths of light were effective in all of the biological responses. The quantitative action spectra, where the amount of light energy needed to give a physiological response is measured at various wavelengths, gives an idea of the absorption spectrum of the receptor pigment. The action spectra show that the peak absorption for promoting effects is quite close to the peak absorption for inhibiting effects (Fig. 11.1). The promotion

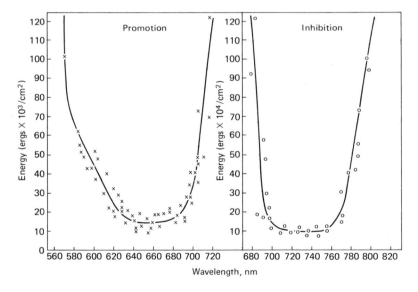

Figure 11.1   Action spectra for promotion and inhibition of anthocyanin formation in sorghum seedlings (var. Wheatland). (From Hendricks, S. B., in *Phytophysiology*, Vol. 1, A. C. Giese, ed., Academic Press, N. Y., 1964, p. 305.)

and inhibition effects are freely reversible. The peak position in the action spectrum and the photoreversibility are suggestive of the possibility that there is a single pigment that is photoisomerized from one form to another. The receptor is called phytochrome and action spectra implicate it in the growth of single cell algae, in *Marchantia,* and in fern spore germination as well as in many higher plant processes. In higher plants red light promotes and far red light inhibits:

Degradation of starch
Anthocyanin formation
Flavonoid biosynthesis
Carotenoid biosynthesis
Protochlorophyllide biosynthesis
Appearance of NADP-linked glyceraldehyde phosphate dehydrogenase
Appearance of certain amino acid activating enzymes
Appearance of inorganic pyrophosphatase activity

and many other processes. At the physiological level, auxin-induced lateral root formation in pea is inhibited by red light, but the same process in *Vignia* is promoted by red light. Plant growth may be either promoted or inhibited by red light depending on the age of the tissue, the species, etc. Red light inhibits coleoptile growth in intact *Oryza* and *Avena* but stimulates growth of coleoptile sections from these two plants. These observations are representative of a formidable catalog of responses and imply that changing the state of the photoreceptor modifies a metabolic system that is poised one way or another toward the final physiological result.

The energy requirement is very low − 1 to 100 k ergs/cm^2 as measured in a bioassay and most of this energy is filtered out before it gets to the phytochrome. Below saturation but above a critical minimum, the response to red light depends linearly on the log of incident energy.

11.2   TIME AND THE PHOTORESPONSE

The time lag in physiological response to illumination should give a hint of the complexity of what goes on between seeing the light and doing something about it. Anthocyanin formation in the reddening of green apples occurs within six hours after irradiation, and this is about the duration of lag in many of the presently measured biochemical responses. An amino acid ammonium lyase that might participate in flavonoid biosynthesis is seen to increase in activity in less than an hour after phytochrome activation of etiolated peas. There is a change in *Mimosa* leaf movement response within five minutes of phytochrome photoconversion. A very fast phytochrome response has been found by Tanada[2] which is related to net charge on the cell surface. Excised root tip sections from dark grown bean seedlings adhere to glass when irradiated with red light and fall away from the glass when irradiated with far red light. The response time is about two minutes. Tanada and, independently, M. J. Jaffe[3] have found that the root tip sections show an electrical potential change from +1 mv in red light to −1 mv in far red light. These phenomena are repeatedly reversible, and the kinetics of glass adhesion-release are the same as the electrical potential shifts. Photoconversion of phytochrome may change the permeability characteristics of cell membranes resulting in an induced electrochemical gradient. This membrane response is suggested as evidence for localizing phytochrome in the cell membrane, but this is a tenuous basis for localization.

The separation of the red-far red treatment by sufficient time allows escape from reversible phytochrome control. The escape is temperature dependent and takes from seconds to hours depending on the species and the process. This escape time is the period in which the metabolic system becomes irreversibly committed.

In the short-term sense, the photoreaction is reversible as rapidly as light can be supplied — switching from promotion to inhibition conditions and back again with the speed of light. In both directions, the photopoising is temperature independent, again arguing for photoconversion between two forms of the same molecule.

11.3   MOLECULAR CHANGES RELATED TO PHYTOCHROME

Butler was able to demonstrate the photoconversion of a pigment *in situ*. He first built an extremely sensitive spectrophotometer to detect very small changes in very dense light scattering material. Butler then found that irradiation of etiolated oat seedlings with 660 nm light caused a decrease in light absorbance at 660 nm and an increase in absorbance at 725 nm. Using 725 nm light to irradiate the tissue gave a decrease in absorbance at 725 nm and an increase at 660 nm. Next, these changes were detected in aqueous extracts of plant material. Difference spectra for the two forms of phytochrome are shown in Figure 11.2. The reaction is described as follows:

$$\text{Red absorbing form (Pr)} \underset{\text{far red light}}{\overset{\text{red light}}{\rightleftharpoons}} \text{far red absorbing form (Pfr)}$$

The only cell-free assay for phytochrome is the reversible photoconversion, and with an appropriately sensitive spectrophotometer one can set out to isolate phytochrome. There have been several attempts to use a microbeam of light to both irradiate and measure phytochrome in specific parts of the living cell. A microbeam can be used to scan the spectra of various parts of the cell; then the cell is irradiated to photoconvert phytochrome and scanned again to detect changes in absorption spectra of the various cell parts.[4] On examining both oat and pea sections with microbeam scanning, Galston found phytochrome in the

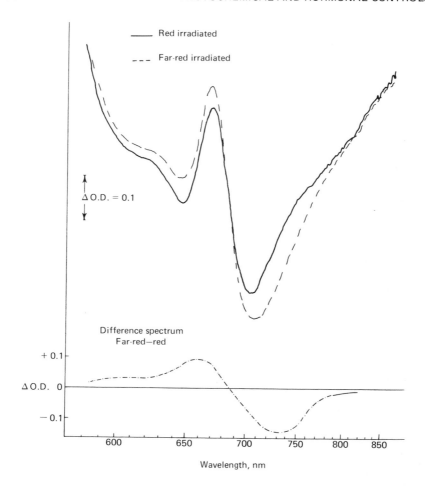

Figure 11.2   Spectra of maize shoots after red and far-red irradiation showing phytochrome photoconversion. [From Butler, W. et al.,
Proc. Natl. Acad. Sci. USA, **45**, 1705 (1959).]

nuclei. Haupt has found evidence for the location of phytochrome in the
chloroplast membranes in the alga *Mougeotia.*[5] An immunochemical localization
of phytochrome indicates that it is in the nucleus, plastids, and cytoplasm.[6]
Perhaps there is some in all membranes. Most of the phytochrome is found with
the soluble protein in leaf homogenates, although a small portion appears to be
tightly bound to membrane debris.[7]

   Isolation of phytochrome is popularly done with etiolated oat or rye. Some
crude extracts contain Pfr "killer" that destroys the reversible absorbance

change. The red absorbing form is three to four times more sensitive than the far red absorbing form to inhibition of absorbance change by glutaraldehyde.[8] Purification of phytochrome is accomplished by ammonium sulfate fractionation, chromatography on calcium phosphate gel, DEAE cellulose, Sephadex, and by disc gel electrophoresis. Spectra of a purified phytochrome preparation are shown in Figure 11.3. The product is 750 fold purified and of high but uncertain

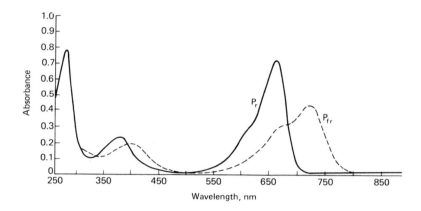

Figure 11.3   Absorption spectra of 750-fold purified *Avena* phytochrome.
Solid curve shows absorbancy after far-red irradiation and
dashed curve after red. [From Mumford, F. E., and Jenner,
E. L., *Biochem.*, 5, 3657 (1966).]

molecular weight. Much of the cell-free work has been plagued by protease degradation of phytochrome.[9] This results in multiple molecular forms that are still photoreactive but of dubious value. Briggs and his associates have prepared a phytochrome of molecular weight 120,000 by avoiding protease attacks.[10] Protease digestion has been turned to advantage in the isolation of a small peptide fragment with the chromophore attached.[11] This should promptly lead to an amino acid sequence around the photoactive site and a knowledge of the mode of chromophore attachment to the protein.

The *in vivo* and *in vitro* action spectra for photoconversion agree, and the quantum efficiency for conversion of the purified phytochrome is approximately 0.7. The *in vitro* photoconversions are independent of temperature down to $-20°$, and one can see a partial bleaching of Pfr at $-196°C$. This partial bleaching is reversible on warming to $-79°C$. These temperature effects suggest a photochemical reaction of the chromophore with attendant changes in the protein structure which are frozen out at low temperature. Since denaturation eliminates photoresponses by phytochrome, the protein is certainly important to the chromophore. Phytochrome shows changes in its circular dichroism spectrum on photoconversion indicating a modest conformational change in the protein.[12] Phytochrome will revert from the Pfr to the Pr form in darkness. *In vitro*, this reversion is temperature sensitive and can be facilitated by reducing agents such as NADH or reduced ferredoxin.[13,14]

The chromophore has been split from the protein by refluxing in acid alcohol indicating that a covalent linkage binds the two together.[15] The isolated pigment is a bilitriene similar but not identical to the chromophore on c-phycocyanin. It is assumed that the photo act causes a cis-trans isomerization with the Pr more stable and perhaps the more planar form. Therefore, one must have a simple isomerization of the chromophore with several conformational intermediates in the transition of the protein as indicated by the low temperature shifts and the several room temperature intermediates recognized by fast absorption changes after a flash of light.[16]

$$Pr \xrightarrow{h\upsilon} Pr^* \longrightarrow r_1 \rightleftarrows \begin{array}{c} r_2 \\ r_3 \\ r_4 \end{array} \longrightarrow Pfr$$

The selective glutaraldehyde inhibition and the claim that Pfr has accessible sulfhydryl groups not seen in Pr afford chemical indications of the change in protein structure accompanying photoconversion.

One must expect minor differences between species, and this is seen in the phytochromes from a green alga and a liverwort.[17] The alga *Mesotaenium* has a phototactic action spectrum implicating phytochrome, and the red absorbance maxima in both the absorption and action spectra are shifted 15 nm to the blue. A phytochrome-like pigment has been found in the blue-green alga *Tolypothrix tenuis*.[18]

How does phytochrome act? Since the pigment is present in catalytic amount, the observed effects must result from some vast amplification of the signal perceived and recorded by phytochrome. Pfr is the form present in the day, and it reverts to Pr in darkness. The Pr form must be present for a critical length of time. If the dark period is interrupted by a brief flash of light, the effect of the dark period, say in allowing short day plants to flower, is nullified. One might imagine a dynamic equilibrium in which the phytochrome must be in the appropriate form for a sufficient length of time to allow sufficient chemical change of something else to initiate the next step in eliciting the physiological response. The phytochrome conversion may induce a whole new pattern of biosynthesis as in the case of the green apple which in darkness converts its acetate to ethanol while a red light treatment induces the conversion of acetate to cyanidin – the red pigment in the apple skin. Since protein synthesis in general and specific enzymes increase after phytochrome activation, it is fashionable to speak of Pfr in gene activation. Actinomycin D inhibits some of the long lag phytochrome responses — red light-induced hook opening in pea and bean, for example. There is a phytochrome mediated repression of lipoxygenase synthesis in mustard seedlings and some controversial evidence on the control of phenylalanine ammonium lyase synthesis by phyto-chrome.[19,20,21] Gene regulation, however, does not seem likely as a mechanism for the rapid leaf movement response of *Mimosa* or the very rapid membrane potential shifts in root tip sections. Here a change in membrane permeability is the immediate effect, and one wonders if phytochrome acts directly or in-directly as a gate to ion flow.

11.4   INDOLEACETIC ACID

The study of plant hormones has been carried out mainly at the physiological level. In most cases, it appears that each hormone elicits a multiplicity of responses.[22] In every case, an exact biochemical knowledge of the mechanism of hormone action is unknown. Indoleacetic acid is perhaps the best known of the plant hormones since its chemical identity and physiological action have been known longest. The biosynthesis and degradation of this hormone proceeds according to the pathway outlined in Figure 11.4. A large variety of analogs of

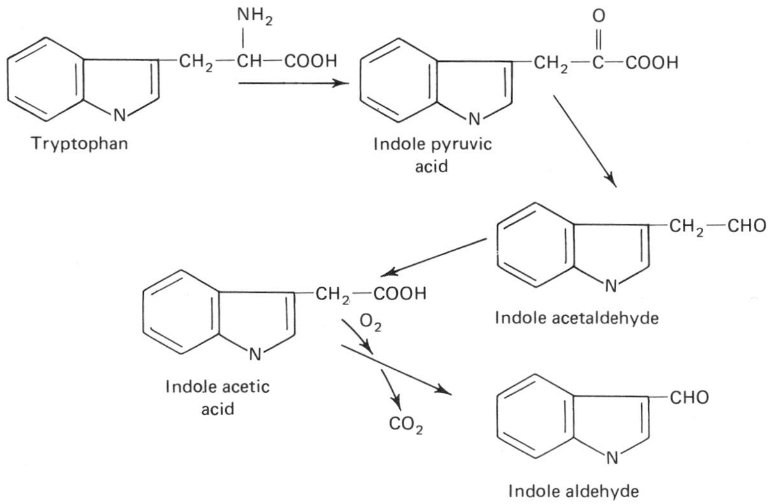

Figure 11.4   Biosynthesis and degradation of the hormone indoleacetic acid.

indoleacetic acid have been synthesized, and many are physiologically active. This hormone functions in all higher plants and possibly in some algae. The physiological concentration of the hormone is in the range of 0.001 to 0.01 mg per kg tissue or 0.01 to 0.1 part per million. Physiological concentrations of the hormone are detected in either the *Avena* curvature test or the pea stem elongation test. In the *Avena* assay, an oat coleoptile is decapitated, and a block of agar containing the hormone is placed asymmetrically on the newly exposed stem tip. Diffusion of the hormone down one side of the stem results in elongation of that side and a bending of the stem on the elongating side. The extent of bending is linear within appropriate limits of hormone concentration (Fig. 11.5). In the pea stem test, segments of the stem are floated in water

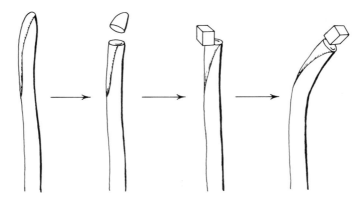

Figure 11.5   A bioassay for indoleacetic acid. The tip of a young oat shoot is removed and an agar block containing the growth hormone is placed on one side of the cut surface. As the hormone diffuses down one side of the shoot, it stimulates growth on that side which bends the plant.

containing hormone, and the increase in length is proportional to the amount of hormone present. Chemical measurement of relatively large amounts of indoleacetic acid is done after either selective extraction or paper chromatographic separation of the hormone; then iron is allowed to form a yellow complex with the indole ring in the Smulchowski test. Indoleacetic acid is believed to act in promoting cell elongation by loosening some polysaccharide links in the cell wall. It is suggested that the hormone might derepress some DNA or modulate some tRNA to induce formation of a wall plasticizing enzyme. In addition to effects on cell wall plasticity, IAA is reported to stimulate the synthesis of ribosomal RNA precursors and to increase the production of cyclic AMP.[23,24,25] The unilateral photodestruction of indoleacetic acid by sunlight in a growing stem is suggested as the basis for phototropism. The hormone is used commercially to control the flowering of ornamental and crop plants, to induce rooting, and to control fruit production.

11.5   ABSCISIC ACID

Abscisic acid arises from mevalonic as has been demonstrated with isotopic tracer and presumably is formed through farnesyl pyrophosphate as outlined in

Figure 11.6.[26,27] Inactivation is accomplished by oxidation of one of the methyl groups attached to the ring to form a methanolic group that can then be glucosylated. Abscisic acid is found in all of the higher plants down to the Pteridophytes and is usually present at a level of 1 mg per 1 to 100 kilograms

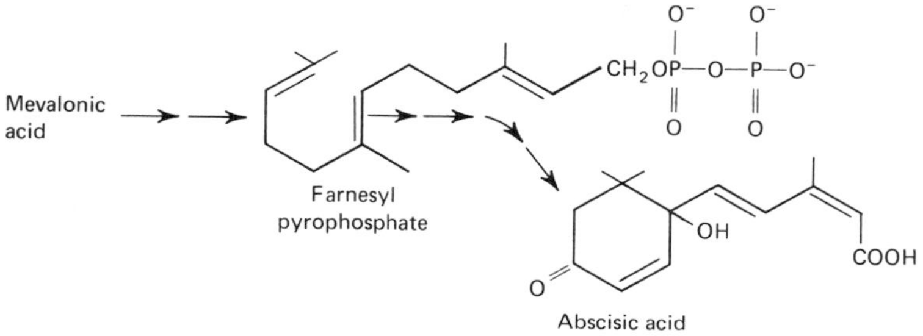

Figure 11.6   The biosynthetic origin of abscisic acid.

(0.01 to 1 ppm). Bioassay of the hormone is by observation of the abscission (falling off) of petiole stalks from a stem segment of a young cotton plant. A sensitive optical assay is achieved in a spectropolarimeter since abscisic acid has a powerful optical rotation at a characteristic wavelength. As to its mechanism of action, abscisic acid generally inhibits both cell division and cell extension. Abscisic acid blocks the effects of growth promoting hormones – gibberellic acid and cytokinins – but the lack of evidence for any specific interactions suggests that the effects of the two types of hormones are independent of each other. Abscisic acid inhibits DNA and RNA synthesis in duckweed and blocks the synthesis of all classes of RNA in radish leaf discs. If abscisic acid is added to a leaf homogenate during chromatin isolation, the chromatin directed RNA synthesis is diminished. Abscisic acid might act as a general gene repressor or might block the activity of RNA polymerase in chromatin. At the physiological level, abscisic acid induces dormancy, induces senescence of excised leaves (i.e., leaves deprived of a supply of growth hormones), inhibits seed germination, and is found in seed coats.

11.6   GIBBERELLIC ACID

Gibberellic acid is a growth promoting hormone that arises from mevalonic acid via geranylgeranyl pyrophosphate as seen in Figure 11.7.[28] There are many variations on the gibberellin structure in nature, and most are biologically active. Gibberellins are found in all of the higher plants and in some fungi and algae, although the latter two may not use the gibberellins as hormones. Gibberellin is measured by a bioassay in which promotion of stem growth of dwarf species of rice, corn, or pea is observed. This assay is effective at the physiological level of around 0.05 ppm. The mechanism of gibberellic acid action is suggested by the

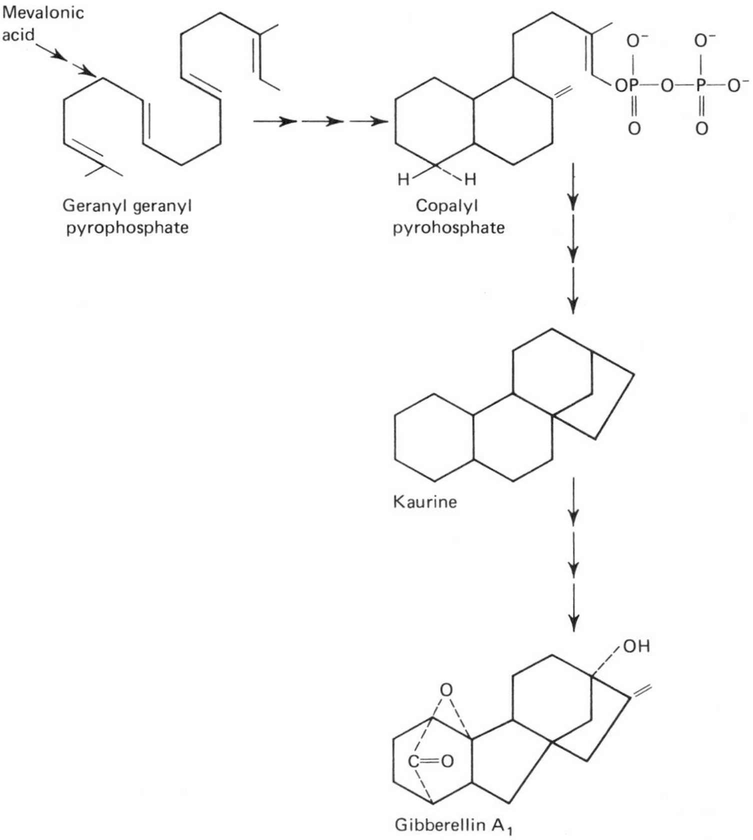

Figure 11.7   The synthesis of gibberellin.

rapid growth of dwarf plants when treated with this hormone. Under these circumstances of rapid growth, since there is certainly a rapid increase in both the DNA and RNA levels in the treated tissue, gibberellic acid may increase nucleic acid synthesis.[29] If one isolates nuclei from dwarf pea in the presence of 10^{-8} M gibberellic acid, these nuclei will incorporate 80% more tritium labelled nucleotides into RNA. Addition of gibberellic acid to the isolated nuclei does not give this effect — the hormone must be present during the cell fractionation.

A dramatic effect of this hormone is the gibberellic acid-induced *de novo* synthesis of hydrolytic enzymes in the barley seed aleurone layer. Imbibition allows gibberellic acid release from the embryo into the aleurone layer where the hydrolases are synthesized to break down seed reserves. How? Since the gibberellic acid is required continuously for a sustained effect, it is not a simple initiator. A very standard complication exists in that abscisic acid will block the gibberellic acid induction of the hydrolase. Again, how? At the gene level? At the level of RNA? Abscisic acid is an antagonist of gibberellins at the physiological level as well as the biochemical level. A cell-free preparation from gibberellic acid treated aleurone cells is reported to synthesize alpha amylase.[30] Gibberellins regulate the synthesis of phospholipids and perhaps in this way control the formation of endoplasmic reticulum in aleurone cells.[31,32] Of special interest is the specific inhibition of gibberellic acid synthesis by some crop controlling agents that dwarf plants and retard growth. AMO 1618 and CCC block the conversion of geranylgeranyl-pyrophosphate to copalyl-pyrophosphate. Phosfon blocks several steps in the pathway including the conversion of copalyl-pyrophosphate to kaurine.

11.7 CYTOKININS

Cytokinins are stimulators of cell division and are chemically identifiable as isopentene adducts of adenine ribosides.[33,34] The structures of several cytokinins are shown in Figure 11.8. The synthesis of cytokinins evidently involves mevalonate and S-adenosyl-methionine as donors of substituents to adenine riboside. The cytokinins are found as regular constituents of tRNA from all species examined, and isotope incorporation data indicate that mevalonate and S-adenosyl-methionine donate the substituents to an adenine riboside resi-

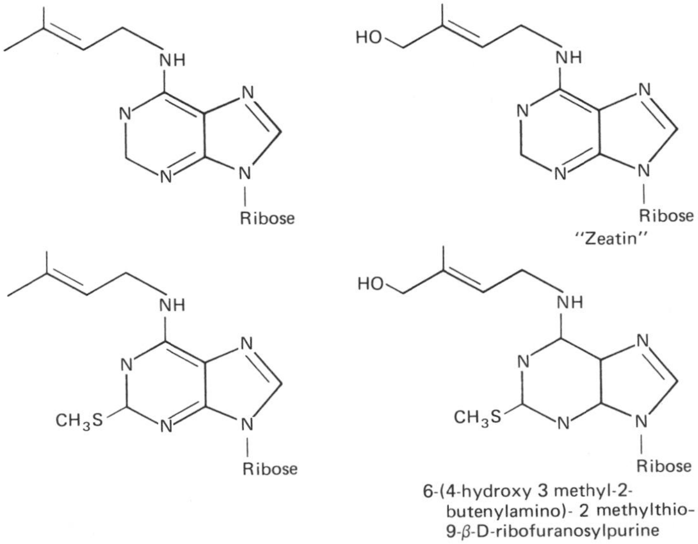

6-(4-hydroxy 3 methyl-2-butenylamino)- 2 methylthio-9-β-D-ribofuranosylpurine

Figure 11.8 Structures of some cytokinins.

due in the tRNA chain. The preformed cytokinins do not appear to be incorporated intact into tRNA. Although compounds with cytokinin activity are found in all tRNAs, cytokinins operating as growth regulators are probably limited to higher plants. Cytokinins are assayed by observing stimulation of cell division in tobacco pith cultures. This assay can detect 5×10^{-11} M zeatin. Some plant cell cultures are not stimulated by addition of cytokinins, presumably because they synthesize an adequate supply. As with all of the plant hormones, there have been many, many structural analogs of the cytokinins synthesized and tested for biological activity. An important conclusion from these analog studies is that cytokinins are in a different class of cell division stimulators from the phenyl ureas that F. C. Stewart had identified as the coconut milk factor needed for carrot explant growth and differentiation in tissue culture. Coconut milk, however, contains a high concentration of zeatin.

The biochemical mechanism of action of the cytokinins as plant hormones is obscure. Cytokinins stimulate DNA synthesis and increase nuclear and nucleolar RNA in onion. The hormone appears to delay protein and nucleic acid loss in senescing tissue and might be supposed to inhibit or repress RNAase and/or DNAase. Cytokinin response can be associated with the appearance of many enzymes in growing tissue, but this could be either a direct causal relationship or a remote secondary effect. It is reported that cytokinin allows better synthesis of RNA by chromatin if the hormone is included in the chromatin isolation media. Cytokinins induce changes in the amounts of two species of leucine tRNA in soybean hypocotyls and cotyledons, but like the enzyme changes mentioned earlier, this could be either a cause or an effect of the cytokinin response.

There is considerable interest in cytokinin as a participant in tRNA (a role that is evidently quite independent of the hormonal effect of this compound).[35,36] In *E. coli* tRNA, cytokinin must be present adjacent to the anticodon of tyrosine tRNA for proper function. Chemical modification of the isopentenyl residue indicates that the cytokinin is not required for amino acid charging but is needed for proper binding of the charged tRNA to the ribosome. In addition to the kinetins within tRNA, there is a report that kinetins bind to plant ribosomes.[37]

In addition to the cytokinins and phenyl urea, a new factor has been reported to stimulate cell division in crown gall tissue. This factor is identified as a glucoside of 3,7 alkyl-2 alkylthio-6-purinone.[38]

11.8   ETHYLENE

Ethylene has long been recognized as a plant growth regulator, but only in relatively recent times has its production by plant tissue and thus its identity as a hormone been verified.[39,40] The existence and effectiveness of ethylene were surmised by a ship's captain who realized that his cargoes of bananas produced volatile material that hastened the ripening of the bananas.

The biosynthetic source of ethylene begins at methionine.[41,42] The synthetic route has been studied in pea seedlings and cauliflower florets. Figure 11.9 indicates some of the possible steps in this pathway.

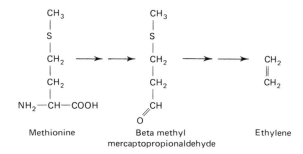

Figure 11.9   A suggested biosynthetic origin for ethylene.

Many unsaturated hydrocarbons mimic the effects of ethylene but with far lower efficiency. It appears that ethylene functions as a hormone in all higher plants. The hormone is bioassayed by the "triple response" of legume seedlings involving the following characteristic morphological changes: (1) reduced elongation, (2) increased radial expansion, and (3) horizontal orientation. A physical assay of very high sensitivity is obtained using vapor-phase chromatography. The resolution of some commercial detectors for vapor-phase chromatograms is quite adequate for measuring physiological concentrations of this hormone.

The biological action of ethylene is most evident in the observation that mature but unripe fruit will produce ethylene, and this ethylene induces ripening. Ethylene induces sprouting in stored potatoes and is used commercially to achieve uniform ripening of bananas, green oranges, and honeydew melons. Indoleacetic acid treatment can cause ethylene production in some tissues, and there is evidence of at least indirect phytochrome control of ethylene production. It is suggested that ethylene stimulates nucleic acid turnover (ethylene treated tissue yields chromatin with about double the RNA synthesis activity of control tissue) or that it affects membrane permeability (ethylene stimulates mitochondrial swelling and enhances respiration of mitochondria in cell-free situations). High concentrations of ethylene enhance the synthesis of phenylalanine ammonium lyase in pea seedlings.[43] As with all of the hormones, there are many chemical effects but no clear biochemical mechanism.

Synthetic plant growth regulators are of enormous practical importance and pose some fascinating problems in basic understanding. A partial listing of such problems includes:

1. Selective herbicides.

2. Harvest synchronizer chemicals.

3. Dormancy control for storage (high CO_2) and dormancy breaking or vernalization to force new growth.

4. Transpiration control chemicals. Alkene succinic acids cause stomates to close reducing both water loss and respiratory CO_2 loss with a much smaller reduction in photosynthetic CO_2 fixation.[44] The molecular mechanism (alteration of membrane permeability?) is unknown, but the use of these reagents offers a chance at better yields and crop protection during dry periods. There is a reported experiment in which a forest valley was sprayed with alkene succinate and the surface soil dried out less than in the previous summer with improved movement of the surface water and without evident damage to the vegetation.

5. Alteration of plant composition. Simazine markedly increases the protein content of a wide variety of plants.[45] It might prove worthwhile to spray grazing land or even jungle with a non-toxic biodegradable analog of simazine and induce a large increase in leaf protein with a great increase in the nutritive value of the plants. In a similar vein, there is a report that phenyl-phosphonic acid induces the loss of stem protein and gives an increase of leaf and seed protein that would concentrate nutritionally valuable material in the edible parts of the plant.[46]

GENERAL REFERENCES

Abeles, F. B. "Biosynthesis and Mechanism of Action of Ethylene," *Ann. Rev. Plant Physiol.*, **23**, 259 (1972).

Briggs, W. R., and H. V. Rice. "Phytochrome: Chemical and Physical Properties and Mechanism of Action," *Ann. Rev. Plant Physiol.*, **23**, 293 (1972).

Furuya, M. "The Biochemistry and Physiology of Phytochrome," *Progress in Phytochemistry*, **1**, 347 (1968).

Steward, F. C., and A. D. Krikorian. *Plants, Chemicals and Growth*. New York: Academic Press, 1971.

Wilkins, M. B. (ed.). *Physiology of Plant Growth and Development*. New York: McGraw-Hill, 1969.

Unseen and Seen

You do not believe in what you cannot see?
Oxygen? Electricity?
Magnetism? Weight?
Photosynthesis?
Someone who can create
Such commonplace miracles? If all this
Leaves you in unbelief,
Look at a leaf.

DOROTHY LONG

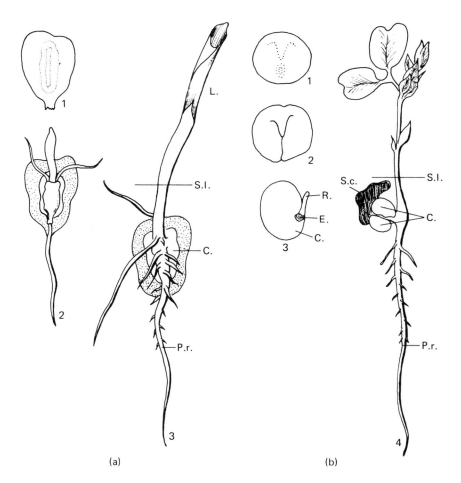

Seeds and seedlings. (a) Corn (*Zea mays*): (1) ungerminated grain; (2) early germination; (3) six-day-old seedling. (b) Garden pea (*Pisum sativum*): (1) seed covered with seed coats, embryonic root barley visible; (2) seed coat removed showing two fleshy cotyledons and embryonic root; (3) hemisected seed; (4) eight-day-old seedling. *Key*: C., cotyledon; E., epicotyl; L., leaf; P.r., primary root; S.I., soil line; S.c., seed coat. (Redrawn from Harold C. Bold, *The Plant Kingdom*, Prentice-Hall, Inc., Englewood Cliffs, N.J., 1961, p. 100.)

GLOSSARY

Abscission. Formation of a special separating layer to sever the leaf or petiole from a deciduous plant.

Aleurone. A layer of cells enclosing the endosperm of the seed. On germination these cells synthesize hydrolytic enzymes to mobilize the nutrient reserves.

Angiosperms. The class of flowering plants having seeds in a closed seed vessel.

Angstrom. A linear unit of 10^{-8} centimeters.

Autotroph. An organism that synthesizes new cellular material from CO_2 using light or an inorganic chemical as its energy source.

Bundle sheath cells. Tough walled cells making up the leaf veins or fibers in grasses. The chloroplast may lack grana.

Callus. A mass of undifferentiated cells.

Cambium. A layer of tissues from which new wood and bark are formed.

Chemosynthetic autotroph. Organism capable of oxidizing an inorganic compound to provide energy and reducing power for the fixation of carbon dioxide.

Chromatophore. Subcellular structure containing the photosynthetic apparatus in bacteria.

Coleoptile. A seed leaf or caplike structure covering epicotyl and immature foliage leaves.

Conifers. A family of plants characterized by needle-shaped leaves, cones, and a resinous wood.

Cotyledon. An embryonic leaf found within the seed.

Dicot. A subclass of flowering plants in which the embryo in the seed has two seed leaves or cotyledons.

Dormancy. A state of minimum metabolic activity.

Endosperm. The nutritive substance within the embryo sac of an ovule or seed.

Etiolated. A plant whitened by lack of photoactivation of chloroplast maturation.

Eucaryote. A cell with a true nucleus.

Gymnosperms. A class of plants whose ovules and seeds are not enclosed in an ovary or case.

Heterotrophic Depending on an external source of reduced carbon for an energy source.

Imbibition. Uptake of water or hydration of a seed prior to germination.

Mesophyll cell. A predominant leaf cell type rich in chloroplasts with stacked membranes or grana.

Monocot. A subclass of flowering plants in which the embryo in the seed has only one seed leaf or cotyledon.

nm (nanometer). One billionth of a meter which is one ten millionth of the distance on the earth's surface from the equator to the pole.

Petiole. Leaf stalk that connects the leaf to the stem.

Pollen. Male gametophytes.

Procaryote. A cell lacking a distinct nucleus surrounded by a nuclear membrane, e.g., bacteria and blue-green algae.

Proplastids. Structures in the etiolated plant which become chloroplasts on greening.

Senescence. Growing old.

Transpiration. Movement of water from the roots to the leaves and its evaporation into the atmosphere.

Variegated. Varied in appearance; a leaf having patches of green and patches of colorless tissue.

AN INFORMAL GLOSSARY OF PLANTS USED IN BIOCHEMICAL RESEARCH

Acer: maple.

Acetabularia: A giant unicellular green alga of special use in microsurgery.

Aesculus: horse chestnut.

Alfalfa: *Medicago sativa.*

Allium: onion.

Anabaena variabilis: a blue-green alga.

Anacystis nidulans: a blue-green alga.

Apple: *Malus.*

Apricot: *Prunus armeniaca.*

Aracea: a family of plants (order Arales) chiefly of tropical distribution distinguished by having the flowers on a fleshy spadix subtended by a leafy spathe.

Arachis hypogaea: peanut.

Artichoke, Jerusalem: *Helianthus tuberosus.*

Arum: wild ginger tuber-bearing low herbs, of few species, in Europe and Western Asia.

Astragulus: a genus that includes seleneum accumulating plants.

Avena sativa: oat.

Avocado: *Perse americana.*

Banana: *Musa.*

Barley: *Hordeum vulgare.*

Bean: *Canavalia, Phaseolus, Vicia, Ricinus, Vignia.*

Beta saccharifera: sugar beet.

Beta var. *Cicla* Moq.: spinach beet.

Beta vulgaris: red beet.

Blackberry: *Rubus.*

Black mustard: *Brassica nigra.*

Blueberry: *Vaccinium.*

Brassica campestris: rapeseed.

Brassica napus: turnip.

Brassica nigra: black mustard.

Brassica oleracea var. *botrytis:* cauliflower.

Brassica oleracea var. *capitata:* cabbage.

Brassica oleracea var. *gemmiferra:* Brussels sprouts.

Brassica pekinensis: Chinese cabbage.

Broad bean: *Vicia faba.*

Brussels sprouts: *Brassica oleracea* var. *gemmifera.*

Cabbage: *Brassica oleracea* var. *capitata.*

Cannabis sativa: hemp.

Cantaloupe: *Cucumis melo.*

Carnation: *Dianthus caryophyllus.*

Carrot: *Daucus carota.*

Carthamus tinctorius: safflower.

Castor bean: *Ricinus communis.*

Cauliflower: *Brassica oleracea* var. *botrytis.*

Chicory: *Chicorium intybus.*

Chinese cabbage: *Brassica pekinensis* or *B. chinensis,* also called celery cabbage.

Chlamydomonas: a genus of green algae.

Chlorella: a genus of green algae.

Chlorobium thiosulfatophilum: a green sulfur photosynthetic bacterium.

Chloropseudomonas ethylicum: a green sulfur photosynthetic bacterium.

Chromatium: a purple sulfur photosynthetic bacterium.

Cichorium endivium: endive.

Cichorium intybus: chicory.

Citrullus lanatus: watermelon.

Citrus aurantifolia: lime.

Citrus limon: lemon.

Citrus paradisi: grapefruit.

Citrus sinensis: orange.

Cocklebur: *Xanthium.*

Coconut: *Cocos nucifera.*

Cocos nucifera: coconut.

Corn: *Zea mays.*

Cotton: *Gossypium.*

Cranberry: *Vaccinium.*

Cucumber: *Cucumis sativus.*

Cucumis melo: cantaloupe, honeydew melon.

Cucumis sativus: cucumber.

Cucurbita: cucurbits, squash, pumpkin.

Cyanidium caldarium: a green alga that contains phycobiliproteins instead of chlorophyll b.

Daucus carota: carrot.

Dianthus caryophyllus: carnation.

Digitalis purpurea: foxglove.

Duckweed: *Lemna.*

Endive: *Cichorium endivium.*

Euglena: a genus of green algae.

Euphorbia pulcherrima: poinsettia.

Flax: *Linum usitatissum.*

Foxglove: *Digitalis purpurea.*
Fragaria: strawberry.
Glycine max: soybean.
Gossypium: cotton.
Gramineae: grasses.
Grapefruit: *Citrus paradisi.*
Helianthus: sunflower.
Helianthus tuberosus: Jerusalem artichoke.
Hemp: *Cannabis sativa.*
Henbane: *Hyoscyamus.*
Hordeum vulgare: barley.
Hydrogenomonas: a chemosynthetic bacterium.
Hyoscyamus: henbane.
Ilex: holly.
Ipomoea: morning glory.
Ipomoea batatas: sweet potato.
Jerusalem artichoke: *Helianthus tuberosus.*
Latuca sativa: lettuce.
Lemna: duckweed.
Lemon: *Citrus limon.*
Lettuce: *Lactuca sativa.*
Lime: *Citrus aurantifolia.*
Linum usitatissimum: flax.
Lupine: *Lupinus.*
Lupinus: lupine.
Lycopersicon esculentum: tomato.
Maize: *Zea mays*, corn.
Malus: apple.
Mango: *Manifera indica.*
Maple: *Acer.*
Marchantia: a liverwort or primitive land plant.
Medicago sativa: alfalfa.
Melilotus albus: sweet clover.
Mentha piperita: mint.
Mesotaenium: a genus of green algae.
Mint: *Mentha piperita.*
Monostroma nitidum: a green alga.
Mougeotia: a genus of green algae.
Mung bean: *Phaseolus aureus.*
Musa: banana.
Navicula: a genus of green algae.
Nicotiana: tobacco.
Oak: *Quercus.*

Oat: *Avena sativa.*
Onion: *Allium.*
Orange: *Citrus sinensis.*
Oryza sativa: rice.
Oxalis: sorrel.
Parsley: *Petroselinum crispum.*
Pea: *Pisum sativum.*
Peanut: *Arachis hypogaea.*
Pear: *Pyrus communis.*
Persea americana: avocado.
Petroselinum crispum: parsley.
Pharbitis nil: Japanese morning glory.
Phaseolus: bean.
Phaseolus aureus: mung bean.
Phaseolus coccineus: scarlet runner bean.
Phaseolus vulgaris: black valentine bean.
Phleum pratense: timothy.
Pinto bean: *Phaseolus vulgaris.*
Pisum sativum: pea.
Platanus occidentalis: sycamore.
Poinsettia: *Euphorbia pulcherrima.*
Porphyra tenera: a red alga.
Potato: *Solanum tuberosum.*
Prunus armeniaca: apricot.
Prunus avium: sweet cherry.
Pumpkin: *Cucurbita pepo.*
Pyrus communis: pear.
Quercus: oak.
Radish: *Raphanus sativus.*
Rapeseed: *Brassica campestris.*
Raphanus sativus: radish.
Red clover: *Trifolium pratense.*
Rhodomicrobium vanneilli: a non-sulfur purple photosynthetic bacterium.
Rhodopseudomonas palustris: a non-sulfur purple photosynthetic bacterium.
Rhodospirillum rubrum: a nonsulfur purple photosynthetic bacterium.
Rice: *Oryza sativa.*
Ricinus communis: castor bean.
Rye: *Secale cereale.*
Saccharum officinarum: sugar cane.
Safflower: *Carthamus tinctorius.*

Scarlet runner bean: *Phaseolus coccineus.*

Scenedesmus: a genus of green algae.

Secale cereale: rye.

Skunk cabbage: *Symplocarpus foetidus.*

Solanum tuberosum: potato.

Soybean: *Glycine max.*

Spinach: *Spinacia oleracea.*

Spinacia oleracea: spinach.

Spinach beet: *Beta* var. *Cicla*, Moq.

Squash: *Cucurbita pepo.*

Strawberry: *Fragaria.*

Sugar beet: *Beta saccharifera*

Sugar cane: *Saccharum officinarum.*

Sunflower: *Helianthus.*

Sweet cherry: *Prunus avium.*

Sweet clover: *Melilotus albus.*

Sweet potato: *Ipomoea batatas.*

Sycamore: *Platanus occidentalis.*

Symplocarpus foetidus: skunk cabbage.

Timothy: *Phleum pratense.*

Tobacco: *Nicotiana.*

Tolypothrix tenuis: a blue-green alga.

Tomato: *Lycopersicon esculentum.*

Trifolium: clover.

Triticum: wheat.

Turnip: *Brassica napus.*

Ulva: a genus of green algae.

Vicia faba: broad bean.

Wheat: *Triticum.*

Xanthium: Cocklebur.

Zea mays: corn, maize.

REFERENCES

CHAPTER 1

1. Bonner, J., and J. E. Varner (eds.). *Plant Biochemistry*. New York: Academic Press, 1965.
2. Kamen, M. D. *Primary Processes in Photosynthesis*. New York: Academic Press, 1963.
3. Unabridged dictionary.
4. Rabinowitch, E. *Photosynthesis and Related Processes*. New York: Wiley (Interscience), 1945.
5. Rabinowitch, E., and Govindjee. *Photosynthesis*. New York: Wiley, 1969.
6. Standard world almanac.
7. Barghoorn, E. S., and S. A. Tyler. *Science*, **147**, 563 (1965).
8. Schopf, J. W., K. A. Kvenvolden, and E. S. Barghoorn. *Proc. Natl. Acad. Sci. USA*, **59**, 639 (1968).
9. Block, K., P. Baronowsky, H. Goldfine, W. J. Lennarz, R. Light, A. T. Norris, and G. Scheuerbrandt. *Fed. Proc.*, **20**, 921 (1961).

10. Dayhoff, M. O. *Atlas of Protein Sequence and Structure.* Silver Springs, Md.: National Biomedical Research Foundation, 1969.
11. Matsubara, H., T. H. Jukes, and C. R. Cantor. *Structure, Function and Evolution of Proteins,* Brookhaven Symposium in Biology, Number 21, 1969, p. 201.
12. Benson, A. M., and K. T. Yasunobu. *J. Biol. Chem.,* **244**, 955 (1969).
13. _____ . *Proc. Natl. Acad. Sci. USA,* **63**, 1269 (1969).
14. Dus, K., K. Sletten, and M. D. Kamen. *J. Biol. Chem.,* **243**, 5507 (1968).
15. Ramshaw, J. A., M. Richardson, and D. Boulter. *Eur. J. Biochem.,* **23**, 475 (1971).
16. Green, B. R., *Biochim. Biophys. Acta,* **254**, 402 (1971).
17. Bendich, A. J., and E. T. Bolton. *Plant Physiol.,* **42**, 959 (1967).

CHAPTER 2

1. Korb, M. J., and H. Beevers. *Plant Physiol.,* **47**, 48 (1971).
2. Dennis, D. T., and T. P. Coultate. *Biochem. Biophys. Res. Commun.,* **2**, 187 (1966).
3. _____ . *Biochim. Biophys. Acta,* **146**, 129 (1967).
4. Kelly, G. J., and J. F. Turner. *Biochim. Biophys. Acta,* **208**, 360 (1970).
5. Bianchetti, R., and M. L. Sartirana. *Biochem. Biophys. Res. Commun.,* **27**, 378 (1967).
6. Scala, J., C. Patrick, and G. Macbeth. *Phytochem.,* **8**, 37 (1969).
7. Scala, J., G. Ketner, and W. H. Jyung. *Arch. Biochem. Biophys.,* **131**, 111 (1969).
8. Rutter, W. J. *Fed. Proc.,* **23**, 1248 (1964).
9. Willard, J. M., and M. Gibbs. *Plant Physiol.,* **43**, 793 (1968).
10. _____ . *Biochim. Biophys. Acta,* **151**, 438 (1968).
11. Russell, G. K., and M. Gibbs. *Biochim. Biophys. Acta,* **132**, 145 (1967).
12. Brooks, K., and R. S. Criddle. *Arch. Biochem. Biophys.,* **117**, 650 (1966).
13. Rapoport, G., L. Davis, and B. L. Horecker. *Arch. Biochem. Biophys.,* **132**, 286 (1969).
14. Takeo, K. *Phytochem.,* **8**, 2127 (1969).
15. Fuller, R. C., and M. Gibbs. *Plant Physiol.,* **34**, 324 (1959).
16. Hageman, R. H., and D. I. Arnon. *Arch. Biochem. Biophys.,* **57**, 421 (1955).
17. Heber, U., N. G. Pon, and M. Heber. *Plant Physiol.,* **38**, 355 (1963).
18. Ogren, W. L., and D. W. Krogmann. *J. Biol. Chem.,* **240**, 4603 (1965).
19. Marcus, A. *Plant Physiol.,* **35**, 126 (1960).
20. Margulies, M. M. *Plant Physiol.,* **40**, 57 (1965).
21. Ziegler, H., and I. Ziegler. *Planta,* **69**, 111 (1966).
22. Melandri, B. A., P. Pupillo, and A. Baccarini Melandri. *Biochim. Biophys. Acta,* **220**, 178 (1970).
23. Ziegler, H., and I. Ziegler. *Planta,* **65**, 369 (1965).
24. Ziegler, H., I. Ziegler, and H. Schmidt-Clausen. *Planta,* **81**, 181 (1968).
25. Ziegler, H., and I. Ziegler. *Biochim. Biophys. Acta,* **126**, 449 (1966).
26. Müller, B. *Biochim. Biophys. Acta,* **205**, 102 (1970).

27. Schulman, M. D., and M. Gibbs. *Plant Physiol.,* **43**, 1805 (1968).
28. Yonuschot, G. R., B. J. Ortwerth, and O. J. Koeppe. *J. Biol. Chem.,* **245**, 4193 (1970).
29. Hudock, G. A., and R. C. Fuller. *Plant Physiol.,* **40**, 1205 (1965).
30. Hood, W., and N. G. Carr. *Biochim. Biophys. Acta,* **146**, 309 (1967).
31. Yamanaka, T., and M. D. Kamen. *Biochim. Biophys. Acta,* **131**, 317 (1967).
32. Hudock, G. A., D. B. Mellin, and R. C. Fuller. *Science,* **150**, 776 (1965).
33. Gibbs, M., and H. Beevers. *Plant Physiol.,* **30**, 343 (1955).
34. Fowler, M. W., and T. Ap Rees. *Biochim. Biophys. Acta,* **201**, 33 (1970).
35. Agrawal, P. K., and D. T. Canvin. *Plant Physiol.,* **47**, 672 (1971).
36. Cheung, W. Y., and M. Gibbs. *Plant Physiol.,* **41**, 731 (1966).
37. Yamamoto, Y. *Plant and Cell Physiol.,* **2**, 288 (1961).
38. Cox, F. F., and D. D. Davis. *Biochem. J.,* **105**, 729 (1967).
39. Duggleby, R. G., and D. T. Dennis. *J. Biol. Chem.,* **245**, 3745 (1970).
40. Omran, R. G., and D. T. Dennis. *Plant Physiol.,* **47**, 43 (1971).
41. Poulsen, L. L., and R. T. Wedding. *J. Biol. Chem.,* **245**, 5709 (1970).
42. Ting, I. P., I. W. Sherman, and W. M. Duggar, Jr. *Plant Physiol.,* **41**, 1083 (1966).
43. Mukerji, S. K., and I. P. Ting. *Arch. Biochem. Biophys.,* **131**, 336 (1969).
44. Ting, I. P., and V. Rocha. *Arch. Biochem. Biophys.,* **147**, 156 (1971).
45. Grimwood, B. G., and R. G. McDaniel. *Biochim. Biophys. Acta,* **220**, 410 (1970).
46. Johnson, H. S., and M. D. Hatch. *Biochem. J.,* **119**, 273 (1970).
47. Johnson, H. S. *Biochem. Biophys. Res. Commun.,* **43**, 703 (1971).
48. Rocha, V., and I. P. Ting. *Arch. Biochem. Biophys.,* **147**, 114 (1971).
49. Cooper, T. G., and H. Beevers. *J. Biol. Chem.,* **244**, 3507 (1969).
50. Smith, A. J., J. London, and R. Y. Stanier. *J. Bact.,* **94**, 972 (1967).
51. Pearce, J., and N. G. Carr. *Biochem. J.,* **105**, 45p (1967).
52. Carr, N. G., and J. Pearce, *Biochem. J.,* **99**, 28p (1966).
53. Marsh, H. V., J. M. Galmiche, and M. Gibbs. *Biochemical Dimensions of Photosynthesis.* Detroit, Mich.: Wayne State University Press, 1965, p. 95.
54. Mattoo, A. K., and V. V. Modi. *Biochem. Biophys. Res. Commun.,* **39**, 895 (1970).
55. Mukerji, S. K., and I. P. Ting. *Arch. Biochem. Biophys.,* **143**, 297 (1971).
56. Lowe, J., and C. R. Slack. *Biochim. Biophys. Acta,* **235**, 207 (1971).
57. Ranson, S. L., in *Plant Biochemistry*, J. Bonner and J. E. Varner (eds.). New York: Academic Press, 1965, p. 493.
58. Seal, S. N., and S. P. Sen. *Plant and Cell Physiol.,* **11**, 119 (1970).
59. Breidenbach, R. W., A. Kahn, and H. Beevers. *Plant Physiol.,* **43**, 705 (1968).
60. Tolbert, N. E. *Ann. Rev. Plant Physiol.,* **22**, 45 (1971).
61. McGregor, D. I., and H. Beevers. *Plant Physiol.,* **44** (Suppl.), 33 (1969).
62. Trelease, R. N., W. M. Becker, P. J. Gruber, and E. H. Newcomb. *Plant Physiol.,* **48**, 461 (1971).
63. Mohapatra, M., E. W. Smith, R. C. Fites, and G. R. Noggle. *Biochem. Biophys. Res. Commun.,* **40**, 1253 (1970).
64. Axelrod, B., and H. Beevers. *Biochim. Biophys. Acta,* **256**, 175 (1972).

CHAPTER 3

1. Bonner, W. D., Jr., in *Plant Biochemistry,* J. Bonner and J. E. Varner (eds.). New York: Academic Press, 1965, p. 89.
2. Miller, R. J., and D. E. Koeppe. *Plant Physiol.,* **47**, 832 (1971).
3. Palmer, J. M., and H. C. Passam. *Biochem. J.,* **122**, 16p (1971).
4. Storey, B. T. *Plant Physiol.,* **48**, 493 (1971).
5. Lance, C., and W. D. Bonner, Jr. *Plant Physiol.,* **43**, 756 (1968).
6. Dutton, P. L., and B. T. Storey. *Plant Physiol.,* **47**, 282 (1971).
7. Ikuma, H., and W. D. Bonner, Jr. *Plant Physiol.,* **42**, 1400 (1967).
8. ———— . *Plant Physiol.,* **42**, 1535 (1967).
9. Storey, B. T. *Plant Physiol.,* **48**, 694 (1971).
10. Bonner, W. D., and D. S. Bendall. *Biochem. J.,* **109**, 47p (1968).
11. Bendall, D. S., and W. D. Bonner, Jr. *Plant Physiol.,* **47**, 236 (1971).
12. Schonbaum, G. R., W. D. Bonner, Jr., B. T. Storey, and J. T. Bahr. *Plant Physiol.,* **47**, 124 (1971).
13. Wilson, S. B. *FEBS Letters,* **15**, 49 (1971).
14. Hoch, G., O. H. Owens, and B. Kok. *Arch. Biochem. Biophys.,* **101**, 171 (1963).
15. Smith, A. J., J. London, and R. Y. Stanier. *J. Bact.,* **94**, 972 (1967).
16. Leach, C. K., and N. G. Carr. *Biochem. J.,* **109**, 4p (1968).
17. Horton, A. A. *Biochem. Biophys. Res. Commun.,* **32**, 839 (1968).
18. Leach, C. K., and N. G. Carr, *Biochem. J.,* **112**, 125 (1969).
19. Webster, D. A., and D. P. Hackett, *Plant Physiol.,* **41**, 599 (1966).
20. ———— . *J. Biol. Chem.,* **241**, 3308 (1966).
21. Atkison, M. R., G. Eckermann, M. Grant, and R. N. Robertson. *Proc. Natl. Acad. Sci. USA,* **55**, 560 (1966).
22. Asahi, T., Y. Honda, and I. Uritani. *Arch. Biochem. Biophys.,* **113**, 498 (1966).
23. Sakano, K., and T. Asahi. *Plant and Cell Physiol.,* **12**, 417 (1971).
24. Kanazawa, Y., T. Asahi, and I. Uritani. *Plant and Cell Physiol.,* **8**, 249 (1967).
25. Chapman, J. M., and J. Edelman. *Biochem. J.,* **103**, 68p (1967).
26. Nawa, Y., and T. Asahi. *Plant Physiol.,* **48**, 671 (1971).
27. Wilson, S. B., and W. D. Bonner, Jr. *Plant Physiol.,* **48**, 340 (1971).
28. Wolstenholme, D. R., and N. J. Gross. *Proc. Natl. Acad. Sci. USA,* **61**, 245 (1968).
29. Kolodner, R., and K. K. Tewari. *Fed. Proc.,* **31**, 876 (1972).
30. Swick, R. W., A. K. Rexroth, and J. L. Stange. *J. Biol. Chem.,* **243**, 3581 (1968).
31. Beattie, D. S. *J. Biol. Chem.,* **243**, 4027 (1968).
32. Bandurski, R. S., in *Plant Biochemistry*, J. Bonner and J. E. Varner (eds.). New York: Academic Press, 1965, p. 467.
33. Mercer, E. I., and G. Thomas. *Phytochem.,* **8**, 2281 (1969).
34. Balharry, G. J. E., and D. J. D. Nicholas. *Biochim. Biophys. Acta,* **220**, 513 (1970).

35. Asada, K. *J. Biol. Chem.*, **242**, 3646 (1967).
36. Mayer, A. M. *Plant Physiol.*, **42**, 324 (1967).
37. Asada, K., G. Tamura, and R. S. Bandurski. *Biochem. Biophys. Res. Commun.*, **30**, 554 (1968).
38. _____. *J. Biol. Chem.*, **244**, 4904 (1969).
39. Schmidt, A., and A. Trebst. *Biochim. Biophys. Acta*, **180**, 529 (1969).
40. Hodson, R. C., J. A. Schiff, and J. P. Mather. *Plant Physiol.*, **47**, 306 (1971).
41. Klepper, L., D. Flesher, and R. H. Hageman. *Plant Physiol.*, **48**, 580 (1971).
42. Eaglesham, A. R. J., and E. J. Hewitt. *Biochem. J.*, **122**, 18p (1971).
43. Vega, J. M., J. Herrera, P. J. Aparicio, A. Paneque, and M. Losada. *Plant Physiol.*, **48**, 294 (1971).
44. Notton, B. A., and E. J. Hewitt. *Plant and Cell Physiol.*, **12**, 465 (1971).
45. Losada, M., A. Paneque, P. J. Aparicio, J. M. Vega, J. Cardenas, and J. Herrera. *Biochem. Biophys. Res. Commun.*, **38**, 1009 (1970).
46. Ingle, J., K. W. Joy, and R. H. Hageman. *Biochem. J.*, **100**, 577 (1966).
47. Filner, P. *Biochim. Biophys. Acta*, **118**, 299 (1966).
48. Heimer, Y. M., and P. Filner. *Biochim. Biophys. Acta*, **230**, 362 (1971).
49. Lips, S. H., and N. Roth-Bejerano. *Science*, **166**, 109 (1969).
50. Travis, R. L., and J. L. Key. *Plant Physiol.*, **48**, 617 (1971).
51. Joy, K. W., and R. H. Hageman. *Biochem. J.*, **100**, 263 (1966).
52. Hattori, A., and J. Myers. *Plant Physiol.*, **41**, 1031 (1966).
53. Paneque, A., and M. Losada. *Biochim. Biophys. Acta*, **128**, 202 (1966).
54. Ramirez, J. M., F. F. Del Campo, A. Paneque, and M. Losada. *Biochim. Biophys. Acta*, **118**, 58 (1966).
55. Zumft, W. G., A. Paneque, P. J. Aparicio, and M. Losada. *Biochem. Biophys. Res. Commun.*, **36**, 980 (1969).
56. Zumft, W. G. *Fed. Proc.*, **31**, 478 (1972).
57. Hucklesby, D. P., and E. J. Hewitt. *Biochem. J.*, **119**, 615 (1970).
58. Leach, R. M., and P. R. Kirk. *Biochem. Biophys. Res. Commun.*, **32**, 685 (1968).
59. Ries, S. K., H. Chmiel, D. R. Dilley, and P. Filner. *Proc. Natl. Acad. Sci. USA*, **58**, 526 (1967).

CHAPTER 4

1. Cooper, T. G., D. Filmer, M. Wishnick, and M. D. Lane. *J. Biol. Chem.*, **244**, 1081 (1969).
2. Everson, R. G. *Phytochemistry*, **9**, 25 (1970).
3. Gibbs, M., E. Latzko, R. G. Everson, and W. Cockburn, in *Harvesting the Sun*, A. San Pietro, F. A. Greer, and T. J. Army (eds.). New York: Academic Press, 1967, p. 111.
4. Walker, D. A. *Plant Physiol.*, **40**, 1157 (1965).
5. Jensen, R. G., and J. A. Bassham. *Plant Physiol.*, **41**, Suppl., lvii (1966).
6. Bucke, C., D. A. Walker, and C. W. Baldry. *Biochem. J.*, **101**, 636 (1966).
7. Cockburn, W., C. W. Baldry, and D. A. Walker. *Biochim. Biophys. Acta*, **143**, 606, 614 (1967).

8. Cockburn, W., D. A. Walker, and C. W. Baldry. *Biochem. J.*, **107**, 89 (1968).
9. Stokes, D. M., and D. A. Walker. *Plant Physiol.*, **48**, 163 (1971).
10. Bassham, J. A., M. Kirk, and R. G. Jensen. *Biochim. Biophys. Acta*, **153**, 211 (1968).
11. Bassham, J. A., A. M. El-Badry, M. R. Kirk, H. J. C. Ottenheym, and H. Springer-Lederer. *Biochim. Biophys. Acta*, **223**, 261 (1970).
12. Buchanan, B. B., P. Schürmann, and P. P. Kalberer. *J. Biol. Chem.*, **246**, 5952 (1971).
13. *The New Scientist*, **40** (Dec. 5, 1968), p. 626.
14. Kawashima, N., and S. G. Wildman. *Ann. Rev. Plant Physiol.*, **21**, 325 (1970).
15. Levine, R. P., and R. K. Togasaki. *Proc. Natl. Acad. Sci. USA*, **53**, 987 (1967).
16. Paulsen, J. M., and M. D. Lane. *Biochem.*, **5**, 2350 (1966).
17. Rutner, A. C., and M. D. Lane. *Biochem. Biophys. Res. Commun.*, **28**, 531 (1967).
18. Rutner, A. C. *Biochem. Biophys. Res. Commun.*, **39**, 923 (1970).
19. Sugiyama, T., T. Ito, and T. Akazawa. *Biochem.*, **10**, 3406 (1971).
20. Kawashima, N., and S. G. Wildman. *Biochim. Biophys. Acta*, **229**, 749 (1971).
21. Kawashima, N., S. Y. Kwok, and S. G. Wildman. *Biochim. Biophys. Acta*, **236**, 578 (1971).
22. Kawashima, N. *Biochem. Biophys. Res. Commun.*, **38**, 119 (1970).
23. Andersen, W. R., G. F. Wildner, and R. S. Criddle. *Arch. Biochem. Biophys.*, **137**, 84 (1970).
24. Smillie, R. M., D. Graham, M. R. Dwyer, A. Grieve, and N. F. Tobin. *Biochem. Biophys. Res. Commun.*, **28**, 604 (1967).
25. Criddle, R. S., B. Dau, G. E. Kleinkopf, and R. C. Huffaker. *Biochem. Biophys. Res. Commun.*, **41**, 621 (1970).
26. Ridley, S. M., J. P. Thornber, and J. L. Bailey. *Biochim. Biophys. Acta*, **140**, 62 (1967).
27. Howell, S. H., and E. N. Moudrianakis. *Proc. Natl. Acad. Sci. USA*, **58**, 1261 (1967).
28. Wishnick, M., and M. D. Lane. *J. Biol. Chem.*, **244**, 55 (1969).
29. Wishnick, M., M. D. Lane, M. C. Scrutton, and A. S. Mildvan. *J. Biol. Chem.*, **244**, 5761 (1969).
30. Sugiyama, T., N. Nakayama, and T. Akazawa. *Biochem. Biophys. Res. Commun.*, **30**, 118 (1968).
31. Bassham, J. A., P. Sharp, and I. Morris. *Biochim. Biophys. Acta*, **153**, 898 (1968).
32. Anderson, L. E., G. B. Price, and R. C. Fuller. *Science*, **161**, 482 (1968).
33. Akazawa, T., K. Sato, and T. Sugiyama. *Plant and Cell Physiol.*, **11**, 39 (1970).
34. Akazawa, T., T. Sugiyama, and H. Kataoka. *Plant and Cell Physiol.*, **11**, 541 (1970).
35. Kuehn, G. D., and B. A. McFadden. *Biochem.*, **8**, 2394, 2403 (1969).

36. Bowes, G., W. L. Ogren, and R. H. Hageman. *Biochem. Biophys. Res. Commun.*, **45**, 716 (1971).
37. Ogren, W. L., and G. Bowes. *Nature, New Biol.*, **230**, 159 (1971).
38. Bowes, G., and W. L. Ogren. *J. Biol. Chem.*, **247**, 2171 (1972).
39. Lorimer, G. H., T. J. Andrews, and N. E. Tolbert. *Fed. Proc.*, **31**, 461 Abs. (1972).
40. Jackson, W. A., and R. J. Volk. *Ann. Rev. Plant Physiol.*, **21**, 385 (1970).
41. Zelitch, I., and P. R. Day. *Plant Physiol.*, **43**, 1838 (1968).
42. Zelitch, I. *Plant Physiol.*, **41**, 1623 (1966).
43. Yu, Y. L., N. E. Tolbert, and G. M. Orth. *Plant Physiol.*, **39**, 643 (1964).
44. Tolbert, N. E. *Ann. Rev. Plant Physiol.*, **22**, 45 (1971).
45. Zelitch, I. *Fed. Proc.*, **31**, 447 Abs. (1972).
46. Trelease, R. N., W. M. Becker, P. J. Gruber, and E. H. Newcomb. *Plant Physiol.*, **48**, 461 (1971).
47. Murray, D. R., and J. W. Bradbeer. *Biochem. J.*, **124**, 69p (1971).
48. Lord, J. M., and M. J. Merrett. *Biochem. J.*, **117**, 929 (1970).
49. Bunt, J. S., and M. A. Heeb. *Biochim. Biophys. Acta*, **226**, 354 (1971).
50. Johnson, H. S., and M. D. Hatch. *Biochem. J.*, **114**, 127 (1969).
51. Hatch, M. D., and C. R. Slack. *Biochem. J.*, **106**, 141 (1968).
52. _____ . *Arch. Biochem. Biophys.*, **120**, 224 (1967).
53. _____ . *Biochem. J.*, **101**, 103 (1966).
54. Slack, C. R. *Biochem. Biophys. Res. Commun.*, **30**, 483 (1968).
55. Hatch, M. D., and C. R. Slack. *Biochem. J.*, **112**, 549 (1969).
56. Butler, L. G., and V. Bennett. *Plant Physiol.*, **44**, 1285 (1969).
57. Lowe, J., and C. R. Slack. *Biochim. Biophys. Acta*, **235**, 207 (1971).
58. Whelan, T., W. M. Sackett, and C. R. Benedict. *Biochem. Biophys. Res. Commun.*, **41**, 1205 (1970).
59. Hatch, M. D. *Biochem. J.*, **125**, 425 (1971).
60. Edwards, G. E., R. Kanai, and C. C. Black. *Biochem. Biophys. Res. Commun.*, **45**, 278 (1971).
61. Evans, M. C. W., B. B. Buchanan, and D. I. Arnon. *Proc. Natl. Acad. Sci. USA*, **55**, 928 (1966).
62. Evans, M. C. W. *Biochem. Biophys. Res. Commun.*, **33**, 146 (1968).
63. Sirevag, R., and J. G. Omerod. *Biochem. J.*, **120**, 399 (1970).
64. Yoch, D. C., and E. S. Lindstrom. *Biochem. Biophys. Res. Commun.*, **28**, 65 (1967).
65. Gehring, U., and D. I. Arnon. *J. Biol. Chem.*, **246**, 4518 (1971).

CHAPTER 5

1. Hassid, W. Z. *Ann. Rev. Plant Physiol.*, **18**, 253 (1967).
2. Axelrod, B., in *Plant Biochemistry*, J. E. Varner and J. Bonner (eds.). New York: Academic Press, 1965, p. 231.
3. Loewus, F. *Ann. Rev. Plant Physiol.*, **22**, 337 (1971).
4. Loewus, M. W., and F. Loewus. *Plant Physiol.*, **48**, 255 (1971).

5. Roberts, R. M. *J. Biol. Chem.*, **246**, 4995 (1971).

6. Wellmann, E., D. Baron, and H. Grisebach. *Biochim. Biophys. Acta*, **244**, 1 (1971).

7. Picken, J. M., and J. Mendicino. *J. Biol. Chem.*, **242**, 1629 (1967).

8. Sandermann, H., Jr., and H. Grisebach. *Eur. J. Biochem.*, **6**, 404 (1968).

9. _____ . *Biochim. Biophys. Acta*, **156**, 435 (1968).

10. Roberts, R. M., R. H. Shah, and F. Loewus. *Plant Physiol.*, **42**, 659 (1967).

11. Wellmann, E., and H. Grisebach. *Biochim. Biophys. Acta*, **235**, 389 (1971).

12. Hawker, J. S. *Biochem. J.*, **105**, 943 (1967).

13. DeFekete, M. A. R. *Eur. J. Biochem.*, **19**, 73 (1971).

14. Slabnik, E., R. B. Frydman, and C. E. Cardini. *Plant Physiol.*, **43**, 1063 (1968).

15. Grimes, W. J., B. L. Jones, and P. Albersheim. *J. Biol. Chem.*, **245**, 188 (1970).

16. Wolosiyk, R. W., and H. G. Pontis. *FEBS Letters*, **16**, 237 (1971).

17. Tanner, W., L. Lehle, and O. Kandler. *Biochem. Biophys. Res. Commun.*, **29**, 166 (1967).

18. Slabnik, E., and R. B. Frydman. *Biochem. Biophys. Res. Commun.*, **38**, 709 (1970).

19. Tsai, C. Y., and O. E. Nelson. *Plant Physiol.*, **44**, 159 (1969).

20. Akazawa, T., in *Plant Biochemistry*, J. E. Varner and J. Bonner (eds.). New York: Academic Press, 1965, p. 258.

21. Hassid, W. Z. *Science*, **165**, 137 (1969).

22. Ozbun, J. L., J. S. Hawker, and J. Preiss. *Plant Physiol.*, **48**, 765 (1971).

23. Tanaka, Y., and T. Akazawa. *Plant and Cell Physiol.*, **12**, 493 (1971).

24. Tsai, C. Y., and O. E. Nelson. *Science*, **151**, 341 (1966).

25. Dickinson, D. B., and J. Preiss. *Plant Physiol.*, **44**, 1058 (1969).

26. Murata, T., N. Nakayama, T. Tanaka, and T. Akazawa. *Arch. Biochem. Biophys.*, **123**, 97 (1968).

27. Perez, C. M., E. P. Palmiano, L. C. Baun, and B. O. Juliano. *Plant Physiol.*, **47**, 404 (1971).

28. Ghosh, H. P., and J. Preiss. *J. Biol. Chem.*, **241**, 4491 (1966).

29. Sanwal, G. G., and J. Preiss. *Arch. Biochem. Biophys.*, **119**, 454 (1967).

30. Sanwal, G. G., E. Greenberg, J. Hardie, E. C. Cameron, and J. Preiss. *Plant Physiol.*, **43**, 417 (1968).

31. Hodges, H. F., R. G. Creech, and J. D. Loerch. *Biochem. Biophys. Acta*, **185**, 70 (1969).

32. Borovsky, D., and W. J. Whelan. *Fed. Proc.*, **31**, 477 Abs. (1972).

33. Varner, J. E., and G. RamChandra. *Proc. Natl. Acad. Sci. USA*, **52**, 100 (1964).

34. Mayer, A. M., and Y. Shain. *Science*, **162**, 1283 (1968).

35. Muarta, T., T. Akazawa, and S. Fukuchi. *Plant Physiol.*, **43**, 1899 (1968).

36. Juliano, B. O., and J. E. Varner. *Plant Physiol.*, **44**, 886 (1969).

37. DeFekete, M. A. R. *Arch. Biochem. Biophys.*, **116**, 368 (1966).

38. Lee, E. Y. C., J. J. Marshall, and W. J. Whelan. *Arch. Biochem. Biophys.*, **143**, 365 (1971).

39. Swain, R. R., and E. E. Dekker. *Biochim. Biophys. Acta*, **122**, 87 (1966).

40. Mayer, A. M., and Y. Shain. *Science*, **162**, 1283 (1968).

41. Ohad, I., I. Friedberg, Z. Neleman, and M. Schramm. *Plant Physiol.*, **47**, 465 (1971).
42. Carr, N. G. *Biochim. Biophys. Acta*, **120**, 308 (1966).
43. Albersheim, P., in *Plant Biochemistry*, J. E. Varner and J. Bonner (eds.). New York: Academic Press, 1965, p. 151.
44. Pinsky, A., and L. Ordin. *Plant and Cell Physiol.*, **10**, 771 (1969).
45. Flowers, H. M., K. K. Batra, J. Kemp, and W. Z. Hassid. *J. Biol. Chem.*, **244**, 4969 (1969).
46. Liu, T. Y., and W. Z. Hassid. *J. Biol. Chem.*, **245**, 1922 (1970).
47. Villemez, C. L. *Biochem. Biophys. Res. Commun.*, **40**, 636 (1970).
48. Kauss, H., A. L. Swanson, and W. Z. Hassid. *Biochem. Biophys. Res. Commun.*, **26**, 234 (1967).
49. McNab, J. M., C. L. Villemez, and P. Albersheim. *Biochem. J.*, **106**, 355 (1968).
50. Lamport, D. T. A. *Ann. Rev. Plant Physiol.*, **21**, 235 (1970).
51. Loewus, F. *Fed. Proc.*, **24**, 855 (1965).
52. ——— . *Ann. Rev. Plant Physiol.*, **22**, 337 (1971).
53. Northcote, D. H., and J. D. Pickett-Heaps. *Biochem. J.*, **98**, 159 (1966).
54. Harris, P. J., and D. H. Northcote. *Biochim. Biophys. Acta*, **237**, 56 (1971).
55. Van der Woude, W. J., C. A. Lembi, and D. J. Morré. *Biochem. Biophys. Res. Commun.*, **46**, 245 (1972).
56. Mühlethaler, K. *Ann. Rev. Plant Physiol.*, **18**, 1 (1967).
57. Newcomb, E. H. *Ann. Rev. Plant Physiol.*, **14**, 43 (1963).
58. Setterfield, G., and S. T. Bayley. *Ann. Rev. Plant Physiol.*, **12**, 35 (1961).
59. Albersheim, P., in *Plant Biochemistry*, J. E. Varner and J. Bonner (eds.). New York: Academic Press, 1965, p. 298.
60. Keegstra, K. G., K. W. Talmadge, W. D. Bauer, and P. Albersheim. *Fed. Proc.*, **31**, 873 Abs. (1972).
61. Freudenberg, K. *Science*, **148**, 595 (1965).
62. Brown, S. A. *Bioscience*, **19**, 115 (1969).
63. Zucker, M. *Biochem. Biophys. Acta*, **208**, 331 (1970).
64. Sadava, D., and M. J. Chrispeels. *Science*, **165**, 299 (1969).
65. Boundy, J. A., J. S. Wall, J. E. Turner, J. H. Woychik, and R. J. Dimler. *J. Biol. Chem.*, **242**, 2410 (1967).
66. Heath, M. F., and D. H. Northcote. *Biochem. J.*, **125**, 953 (1971).
67. Cleland, R. *Plant Physiol.*, **43**, 865 (1968).
68. Chrispeels, M. J. *Plant Physiol.*, **45**, 223 (1970).
69. Sadava, D., and M. J. Chrispeels. *Biochim. Biophys. Acta*, **227**, 278 (1971).
70. Holleman, J. *Proc. Natl. Acad. Sci. USA*, **57**, 50 (1967).
71. Cleland, R., and A. C. Olson. *Biochem.*, **7**, 1745 (1968).
72. ——— . *Biochem.*, **6**, 32 (1967).
73. Cleland, R., and A. M. Karlsnes. *Plant Physiol.*, **42**, 669 (1967).
74. Roberts, R. M., A. B. Connor, and J. J. Cetorelli. *Biochem. J.*, **125**, 999 (1971).
75. Jervis, L., and M. Hallaway. *Biochem. J.*, **117**, 505 (1970).
76. Ruesink, A. W., and K. V. Thimann. *Proc. Natl. Acad. Sci. USA*, **54**, 56 (1965).
77. Ruesink, A. W. *Plant Physiol.*, **47**, 192 (1971).

78. Otsuki, Y., and I. Takebe. *Plant and Cell Physiol.,* **10**, 917 (1969).
79. Pojnar, E., and E. C. Cocking. *Biochem. J.,* **103**, 74p (1967).
80. Withers, L. A., J. B. Power, and E. C. Cocking. *Biochem. J.,* **124**, 47p (1971).
81. Cocking, E. C., *Ann. Rev. Plant Physiol.,* **23**, 29 (1972).

CHAPTER 6

1. Mudd, J. B. *Ann. Rev. Plant Physiol.,* **18**, 229 (1967).
2. Stumpf, P. K., in *Plant Biochemistry.* J. E. Varner and J. Bonner (eds.). New York: Academic Press, 1965, p. 323.
3. Nichols, B. W., and A. T. James, in *Progress in Phytochemistry.* L. Reinhold and Y. Liwschitz (eds.). London: Interscience, 1968, p. 1.
4. Simoni, R. D., R. S. Criddle, and P. K. Stumpf. *J. Biol. Chem.,* **242**, 573 (1968).
5. Matsumura, S., and P. K. Stumpf. *Arch. Biochem. Biophys.,* **125**, 932 (1968).
6. Rinne, R. W., and D. T. Canvin. *Plant and Cell Physiol.,* **12**, 387 (1971).
7. Huang, K. P., and P. K. Stumpf. *Arch. Biochem. Biophys.,* **143**, 412 (1971).
8. Delo, J., M. L. Ernst-Fonberg, and K. Bloch. *Arch. Biochem. Biophys.,* **143**, 384 (1971).
9. Ernst-Fonberg, M. L., and K. Bloch. *Arch. Biochem. Biophys.,* **143**, 392 (1971).
10. Harwood, J. L., and P. K. Stumpf. *Arch. Biochem. Biophys.,* **142**, 281 (1971).
11. Stumpf, P. K., and N. K. Boardman. *J. Biol. Chem.,* **245**, 2579 (1970).
12. Nagi, J., and K. Bloch. *J. Biol. Chem.,* **242**, 357 (1968).
13. _____ . *J. Biol. Chem.,* **243**, 4626 (1968).
14. Vijay, I. K., and P. K. Stumpf. *J. Biol. Chem.,* **246**, 2910 (1971).
15. Gurr, M. I., and P. Brawn. *Eur. J. Biochem.,* **17**, 19 (1970).
16. Roughan, P. G. *Biochem. J.,* **117**, 1 (1970).
17. Morris, L. J. *Biochem. Biophys. Res. Commun.,* **29**, 311 (1967).
18. Galliard, T., and P. K. Stumpf. *J. Biol. Chem.,* **241**, 5806 (1966).
19. Ongun, A., and J. B. Mudd. *J. Biol. Chem.,* **243**, 1558 (1968).
20. Yatsu, L. Y., T. J. Jacks, and T. P. Hensarling. *Plant Physiol.,* **48**, 675 (1971).
21. Kolattukudy, P. E. *Science,* **159**, 498 (1968).
22. _____ . *Ann. Rev. Plant Physiol.,* **21**, 163 (1970).
23. Kolattukudy, P. E., and J. S. Buckner. *Biochem. Biophys. Res. Commun.,* **46**, 801 (1972).
24. Kolattukudy, P. E., T. J. Walton, and R. P. S. Kushawa. *Biochem. Biophys. Res. Commun.,* **42**, 739 (1971).
25. Tavener, R. J. A., and D. L. Laidman. *Biochem. J.,* **109**, 9p (1968).
26. Ory, R. L., L. Y. Yatsu, and H. W. Kircher. *Arch. Biochem. Biophys.,* **123**, 255 (1968).
27. Cooper, T. G., and H. Beevers. *Fed. Proc.,* **28**, 538 (1969).

28. Panter, R. A., and J. B. Mudd. *FEBS Letters,* **5**, 169 (1969).
29. Hutton, D., and P. K. Stumpf. *Arch. Biochem. Biophys.,* **142**, 48 (1971).
30. Hitchcock, C. H. S., and L. J. Morris. *Eur. J. Biochem.,* **17**, 39 (1970).
31. Robinson, T. *The Organic Constituents of Higher Plants.* Minneapolis, Minn.: Burgess Publ., 1963.
32. Porter, J. W., and D. G. Anderson. *Ann. Rev. Plant Physiol.,* **18**, 197 (1967).
33. Goodwin, T. W. *Biochem. J.,* **123**, 293 (1971).
34. Subbarayan, C., S. C. Kushwaha, G. Suzue, and J. W. Porter. *Arch. Biochem. Biophys.,* **137**, 547 (1970).
35. Kushwana, S. C., C. Subbarayan, D. A. Beeler, and J. W. Porter. *J. Biol. Chem.,* **244**, 3635 (1969).
36. Burns, E. R., G. A. Buchanan, and M. C. Carter. *Plant Physiol.,* **47**, 144 (1971).
37. Rees, H. H., L. J. Goad, and T. W. Goodwin. *Biochem. J.,* **107**, 417 (1968).
38. West, C. A. *Biochem. J.,* **114**, 3p (1969).
39. Murphy, A. J., and C. A. West. *Arch. Biochem. Biophys.,* **133**, 395 (1969).
40. Hepper, C. M., and B. G. Audley. *Biochem. J.,* **114**, 379 (1969).
41. Langenheim, J. H. *Science,* **163**, 1157 (1969).
42. Bogorad, L., in *Plant Biochemistry.* J. E. Varner and J. Bonner (eds.). New York: Academic Press, 1965, p. 717.
43. Lascelles, J., *Tetrapyrrole Biosynthesis and Its Regulation.* New York: Benjamin, 1964.
44. Wellburn, F. A. M., and A. R. Wellburn. *Biochem. Biophys. Res. Commun.,* **45**, 747 (1971).
45. Rebeiz, C. A., and P. A. Castelfranco. *Plant Physiol.,* **47**, 24 (1971).
46. Pluscec, J., and L. Bogorad. *Biochem.,* **9**, 4737 (1970).
47. Troxler, R. E., A. Brown, R. Lester, and P. White. *Science,* **167**, 192 (1970).
48. Ellsworth, R. K., and S. Aronoff. *Arch. Biochem. Biophys.,* **125**, 269 (1968).
49. Nandi, D. L., K. F. Baker-Cohen, and D. Shemin. *J. Biol. Chem.* **243**, 1224 (1968).
50. Nandi, D. L., and D. Shemin. *J. Biol. Chem.,* **243**, 1236 (1968).
51. Gassman, M., and L. Bogorad. *Plant Physiol.,* **42**, 774 (1967).
52. Beale, S. I. *Plant Physiol.,* **48**, 316 (1971).
53. Shetty, A. S., and G. W. Miller. *Biochem. J.,* **114**, 331 (1969).
54. Llambias, E. B. C., and A. M. Del C. Battle. *Biochem. J.,* **121**, 327 (1971).
55. Frydman, R. B., and B. Frydman. *Arch. Biochem. Biophys.,* **136**, 193 (1970).
56. Machold, O., and U. W. Stephan. *Phytochem.,* **8**, 2189 (1969).
57. Hsu, W. P., and G. W. Miller. *Biochem. J.,* **117**, 215 (1970).
58. Goldin. B. R., and H. N. Little. *Biochim. Biophys. Acta,* **171**, 321 (1969).
59. Henningsen, K. W., and A. Kahn. *Plant Physiol.,* **47**, 685 (1971).
60. Thorne, S. W. *Biochim. Biophys. Acta,* **226**, 113 (1971).
61. Argyroudi-Akoyunoglou, J. H., Z. Feleki, and G. Akoyunoglou. *Biochem. Biophys. Res. Commun.,* **45**, 606 (1971).
62. Wellburn, A. R. *Phytochem.,* **9**, 2311 (1970).

63. Ellsworth, R. K., and S. Aronoff. *Arch. Biochem. Biophys.,* **125**, 35 (1968).
64. Ellsworth, R. K., H. J. Perkins, J. P. Detwiller, and K. Liu. *Biochim. Biophys. Acta,* **223**, 275 (1970).
65. Rebeiz, C. A., and P. A. Castelfranco. *Plant Physiol.,* **47**, 33 (1971).
66. Troxler, R. F., and A. Brown. *Biochim. Biophys. Acta,* **215**, 503 (1970).

CHAPTER 7

1. Sun, A. S. K., and K. Sauer. *Biochim. Biophys. Acta,* **234**, 399 (1971).
2. Myers, J. *Ann. Rev. Plant Physiol.,* **22**, 289 (1971).
3. Blinks, L. R., in *Photophysiology.* A. C. Giese (ed.). New York: Academic Press, 1964, Vol. I, 199.
4. Duysens, L. N. M., J. Amesz, and B. M. Kamp. *Nature,* **190**, 510 (1961).
5. McSwain, B. D., and D. I. Arnon. *Proc. Natl. Acad. Sci. USA,* **61**, 989 (1968).
6. Sane, P. V., and R. B. Park. *Biochem. Biophys. Res. Commun.,* **44**, 491 (1971).
7. Ben-Hayyim, G., and M. Avron. *Photochem. and Photobiol.,* **14**, 389 (1971).
8. Myers, J., and J. R. Graham. *Plant Physiol.,* **48**, 282 (1971).
9. Hall, D. O., R. Cammack, and K. K. Rao. *Nature,* **233**, 136 (1971).
10. Gibson, J. F., D. O. Hall, J. H. M. Thornley, and F. R. Whatley. *Proc. Natl. Acad. Sci. USA,* **56**, 987 (1966).
11. Evans, M. C. W., D. O. Hall, H. Bothe, and F. R. Whatley. *Biochem. J.,* **109**, 45p (1968).
12. Zanetti, G., and G. Forti. *J. Biol. Chem.,* **241**, 279 (1966).
13. Buchanan, B. B., and M. C. W. Evans. *Biochim. Biophys. Acta,* **180**, 123 (1969).
14. Ogren, W. L., and D. W. Krogmann. *J. Biol. Chem.,* **240**, 4603 (1965).
15. Yamanaka, T., and M. D. Kamen. *Biochim Biophys. Acta,* **131**, 317 (1967).
16. Keister, D. L., and N. J. Yike. *Biochem.,* **6**, 3847 (1967).
17. Huang, K., S.-I. Tu, and J. H. Wanf. *Biochem. Biophys. Res. Commun.,* **34**, 48 (1969).
18. Nelson, N., and J. Neumann. *J. Biol. Chem.,* **244**, 1926, 1932 (1969).
19. Shin, M., and A. San Pietro. *Biochem. Biophys. Res. Commun.,* **33**, 38 (1968).
20. Foust, G. P., S. G. Mayhew, and V. Massey. *J. Biol. Chem.,* **244**, 964 (1969).
21. Fredricks, W. W., and J. M. Gehl. *J. Biol. Chem.,* **246**, 1201 (1971).
22. Nakamura, S., and T. Kimura. *FEBS Letters,* **15**, 352 (1971).
23. Fredricks, W. W., and J. M. Kohlmann. *J. Biol. Chem.,* **244**, 522 (1969).
24. Smillie, R. M. *Biochem. Biophys. Res. Commun.,* **20**, 621 (1965).
25. Trebst, A., and H. Bothe. *Ber. Deut. Bot. Ges.,* **79**, 44 (1966).
26. Zumft, W. G., and H. Spiller. *Biochem. Biophys. Res. Commun.,* **45**, 112 (1971).
27. Heldt, H. W. *FEBS Letters,* **5**, 11 (1969).

28. Heber, U. W., and K. A. Santarius. *Biochim. Biophys. Acta,* **109**, 390 (1965).
29. Stocking, C. R., and S. Larson. *Biochem. Biophys. Res. Commun.,* **37**, 278 (1969).
30. Losada, M., J. M. Ramirez, A. Paneque, and F. F. Del Campo. *Biochim. Biophys. Acta,* **109**, 86 (1965).
·31. Hattori, A., and J. Myers. *Plant Physiol.,* **41**, 1031 (1966).
32. Asada, K., G. Tamura, and R. S. Bandurski. *Biochem. Biophys. Res. Commun.,* **30**, 554 (1968).
33. Evans, M. C. W., B. B. Buchanan, and D. I. Arnon. *Proc. Natl. Acad. Sci. USA,* **55**, 928 (1969).
34. Hiyama, T., and B. Ke. *Proc. Natl. Acad. Sci. USA,* **68**, 1010 (1971).
35. _____ . *Arch Biochem. Biophys.,* **147**, 99 (1971).
36. Malkin, B., and A. J. Bearden. *Proc. Natl. Acad. Sci. USA,* **68**, 16 (1971).
37. Bearden, A. J., and R. Malkin. *Biochem. Biophys. Res. Commun.,* **46**, 1299 (1972).
38. Zweig, G., N. Shavit, and M. Avron. *Biochim. Biophys. Acta,* **109**, 332 (1965).
39. Black. C. C., Jr. *Biochim. Biophys. Acta,* **120**, 332 (1966).
40. Kok, B., H. J. Rurainski, and O. V. H. Owens. *Biochim. Biophys. Acta,* **109**, 347 (1965).
41. Honeycutt, R. C., and D. W. Krogmann. *Biochim. Biophys. Acta,* **197**, 267 (1970).
42. Parson, W. W., *Biochim. Biophys. Acta,* **189**, 384 (1969).
43. Yocum, C. F., and A. San Pietro. *Biochem. Biophys. Res. Commun.,* **36**, 614 (1969).
44. _____ . *Arch. Biochem. Biophys.,* **140**, 152 (1972).
45. Fujita, Y., and J. Myers. *Arch. Biochem. Biophys.,* **112**, 519 (1965).
46. Regitz, G., R. Berzborn, and A. Trebst. *Planta,* **91**, 8 (1970).
47. Kok, B., in *Plant Biochemistry.* J. Bonner and J. E. Varner (eds.). New York: Academic Press, 1965, p. 903.
48. _____ , in *Harvesting the Sun.* M. Greer, T. Army, and A. San Pietro (eds.). New York: Academic Press, 1968, p. 29.
49. Thornber, J. P., C. A. Smith, and J. L. Bailey. *Biochem. J.,* **100**, 14p (1966).
50. Kung, S. D., and J. P. Thornber. *Biochim. Biophys. Acta,* **253**, 285 (1971).
51. Murata, T., F. Toda, K. Uchino, and E. Yakushiji. *Biochim. Biophys. Acta,* **245**, 208 (1971).
52. Thornber, J. P., and J. M. Olsen. *Biochem.,* **7**, 2242 (1968).
53. Kok, B. *Plant Physiol.,* **34**, 184 (1959).
54. Kok, B., and G. Hoch, in *Light and Life.* W. D. McElroy and B. Glass (eds.). Baltimore, Md.: Johns Hopkins University Press, 1961, p. 397.
55. Ke, B., T. Ogawa, T. Hiyama, and L. P. Vernon. *Biochim. Biophys. Acta,* **226**, 53 (1971).
56. Ruuge, E. K., and S. Izawa. *Fed. Proc.,* **31**, 901 Abs. (1972).
57. Weaver, E. *Ann. Rev. Plant Physiol.,* **19**, 283 (1968).
58. Pratt, L. H., and N. I. Bishop. *Biochim. Biophys. Acta,* **153**, 664 (1968).
59. Schleyer, H. *Biochim. Biophys. Acta,* **153**, 427 (1968).

60. Loach, P. A., and D. L. Sekura. *Biochem.*, **7**, 2642 (1968).
61. McElroy, J. D., G. Feher, and D. C. Mauzerall. *Biochim. Biophys. Acta*, **172**, 180 (1969).
62. Ogawa, T., and L. P. Vernon. *Biochim. Biophys. Acta*, **197**, 292 (1970).
63. Reed, D. W., and R. K. Clayton. *Biochim. Biophys. Res. Commun.*, **30**, 471 (1968).

CHAPTER 8

1. Katoh, S., I. Suga, I. Shiratori, and A. Takamiya. *Arch. Biochem. Biophys.*, **94**, 136 (1961).
2. Milne, P. R., and J. R. E. Wells. *J. Biol. Chem.*, **245**, 1566 (1970).
3. Gorman, D. S., and R. P. Levine. *Proc. Natl. Acad. Sci. USA*, **54**, 1665 (1965).
4. Levine, R. P. *Science*, **162**, 768 (1968).
5. Avron, M., and A. Shneyour. *Biochim. Biophys. Acta*, **226**, 498 (1971).
6. Lien, S., and T. T. Bannister. *Biochim. Biophys. Acta*, **245**, 465 (1971).
7. Smillie, R. M., K. S. Andersen, and D. G. Bishop. *FEBS Letters*, **13**, 318 (1971).
8. Lightbody, J. J., and D. W. Krogmann. *Biochim. Biophys. Acta*, **131**, 508 (1967).
9. Hauska, G. A., R. E. McCarty, R. J. Berzborn, and E. Racker. *J. Biol. Chem.*, **246**, 3524 (1971).
10. Brand, J., T. Baszynski, F. L. Crane, and D. W. Krogmann. *J. Biol. Chem.*, **247**, 2814 (1972).
11. Davenport, J., and R. Hill. *Proc. Roy. Soc. B*, **139**, 327 (1952).
12. Bendall, D. S., and R. Hill. *Ann. Rev. Plant Physiol.*, **19**, 167 (1968).
13. Nelson, N. *Fed. Proc.*, **31**, 890 Abs. (1972).
14. Singh, J., and A. R. Wasserman. *J. Biol. Chem.*, **246**, 3532 (1971).
15. Dus, K., K. Sletten, and M. D. Kamen. *J. Biol. Chem.*, **243**, 5507 (1968).
16. Duysens, L. N. M. *Science*, **121**, 210 (1955).
17. Chance, B., D. DeVault, W. W. Hildreth, W. W. Parson, and M. Nishimura. *Brookhaven Symposium No. 19* (1967), p. 115.
18. Parson, W. W. *Biochim. Biophys. Acta*, **153**, 248 (1968).
19. Hildreth, W. W. *Biochim. Biophys. Acta*, **153**, 197 (1968).
20. _____ . *Plant Physiol.*, **43**, 303 (1968).
21. Boardman, N. K., J. M. Anderson, and R. C. Hillier. *Biochim. Biophys. Acta*, **234**, 126 (1971).
22. Bendall, D. S., and D. Sofrova. *Biochim. Biophys. Acta*, **234**, 371 (1971).
23. Katoh, S., and A. San Pietro. *Arch. Biochem. Biophys.*, **118**, 488 (1967).
24. _____ . *Arch. Biochem. Biophys.*, **121**, 211 (1967).
25. Pratt, L. H., and N. I. Bishop. *Biochim. Biophys. Acta*, **153**, 664 (1968).
26. Lightbody, J. J., and D. W. Krogmann. *Biochim. Biophys. Acta*, **131**, 508 (1967).
27. Elstner, E., E. Pistorius, P. Böger, and A. Trebst. *Planta*, **79**, 146 (1968).
28. Bendall, D. S., and R. Hill. *Ann. Rev. Plant Physiol.*, **19**, 167 (1968).

29. Garewal, H. G., J. Singh, and A. R. Wasserman. *Biochem. Biophys. Res. Commun.*, **44**, 1300 (1971).
30. Bendall, D. S. *Biochem. J.*, **109**, 46p (1968).
31. Hind, G., and J. M. Olson. *Ann. Rev. Plant Physiol.*, **19**, 249 (1968).
32. Hind, G. *Biochim. Biophys. Acta,* **153**, 235 (1968).
33. Cramer, W. A., and W. L. Butler. *Biochim. Biophys. Acta,* **143**, 332 (1967).
34. Hildreth, W. W. *Plant Physiol.*, **43**, 303 (1968).
35. Knaff, D. B., and D. I. Arnon. *Proc. Natl. Acad. Sci. USA*, **63**, 956 (1969).
36. Erixon, K., and W. L. Butler. *Biochim. Biophys. Acta,* **234**, 381 (1971).
37. Floyd. R. A., B. Chance, and D. Devault. *Biochim. Biophys. Acta,* **226**, 103 (1971).
38. Hiller, R. G., J. M. Anderson, and N. K. Boardman. *Biochim. Biophys. Acta,* **245**, 439 (1971).
39. Cramer, W. A., and H. Böhme. *Biochim. Biophys. Acta,* **256**, 358 (1972).
40. Levine, R. P., D. S. Gorman, M. Avron, and W. L. Butler. *Brookhaven Symposium in Biology, No. 19* (1967), p. 143.
41. Lynch, V. H., and C. S. French. *Arch. Biochem. Biophys.*, **70**, 382 (1957).
42. Crane, F. L. *Plant Physiol.*, **34**, 128 (1959).
43. Bishop, N. I. *Proc. Natl. Acad. Sci. USA*, **45**, 1696 (1959).
44. Barr, R., and F. L. Crane. *Plant Physiol.*, **42**, 1255 (1967).
45. Barr, R., M. D. Henninger, and F. L. Crane. *Plant Physiol.*, **42**, 1246 (1967).
46. Sun, E., R. Barr, and F. L. Crane. *Plant Physiol.*, **43**, 1935 (1968).
47. Das, B. C., M. Lounasmaa, C. Tendille, and E. Lederer. *Biochem. Biophys. Res. Commun.*, **26**, 211 (1967).
48. Bucke, C., R. M. Leech, M. Hallaway, and R. A. Morton. *Biochim. Biophys. Acta*, **112**, 19 (1966).
49. Redfearn, E., and R. Powls. *Biochem. J.*, **106**, 50p (1968).
50. Powls, R., E. Redfearn, and S. Trippett. *Biochem. Biophys. Res. Commun.*, **33**, 408 (1968).
51. Arnon, D. I., and A. A. Horton. *Acta Chem. Scand.*, **17**, S 135 (1963).
52. Lightbody, J. J., and D. W. Krogmann. *Biochim. Biophys. Acta,* **120**, 57 (1966).
53. Böhme, H., and W. A. Cramer. *FEBS Letters,* **15**, 349 (1971).
54. Magree, L., M. D. Henninger, and F. L. Crane. *J. Biol. Chem.*, **241**, 5197 (1966).
55. Weaver, E. C. *Ann. Rev. Plant Physiol.*, **19**, 283 (1968).
56. Kohl, D. H., and P. M. Wood. *Plant Physiol.*, **44**, 1439 (1969).
57. Kok, B., and E. A. Datko. *Plant Physiol.*, **40**, 1171 (1965).
58. Knaff, D. B., and D. I. Arnon. *Proc. Natl. Acad. Sci. USA*, **63**, 956 (1969).
59. Govindjee, S. Ichimura, C. Cederstrand, and E. Rabinowitch. *Arch. Biochem. Biophys.*, **89**, 322 (1960).
60. Duysens, L. M. N., and J. Sweer, in *Studies on Microalgae and Photosynthetic Bacteria.* Tokyo: University of Tokyo, 1963, p. 353.
61. Kok, B., in *Photosynthetic Processes in Green Plants*, NAS-NRC, (1963), p. 45.
62. Forbush, B., and B. Kok. *Biochim. Biophys. Acta,* **162**, 243 (1968).
63. Kok, B., and G. M. Cheniae, in *Current Topics in Bioenergetics,* D. R. Sanadi (ed.). Vol. 1. New York: Academic Press, 1966.

64. Malkin, S., and B. Kok. *Biochim. Biophys. Acta,* **126,** 413 (1966).
65. Joliot, P. *Brookhaven Symposium in Biology, No. 19* (1967), p. 418.
66. _____ . *Biochim. Biophys. Acta,* **102,** 116 (1962).
67. Joliot, P., A. Joliot, and B. Kok. *Biochim. Biophys. Acta,* **153,** 635 (1968).
68. Bertsch, W., J. R. Azzi, and J. B. Davidson. *Biochim. Biophys. Acta,* **143,** 129 (1967).
69. Jones, L. W. *Proc. Natl. Acad. Sci. USA,* **58,** 75 (1967).
70. Kelley, J., and K. Sauer. *Biochem.,* **7,** 882 (1968).
71. Cheniae, G. M., and I. F. Martin. *Biochem. Biophys. Res. Commun.,* **28,** 89 (1967).
72. Homann, P. H. *Plant Physiol.,* **42,** 997 (1967).
73. Cheniae, G. M., and I. F. Martin. *Biochim. Biophys. Acta,* **153,** 819 (1968).
74. _____ . *Biochim. Biophys. Acta,* **253,** 167 (1971).
75. Jenkins, R., and D. E. Griffiths. *Biochem. J.,* **116,** 40p (1970).
76. Cheniae, G. M., and I. F. Martin. *Biochim. Biophys. Acta,* **197,** 219 (1970).
77. Hind, G., H. Y. Nakatani, and S. Izawa. *Biochim. Biophys. Acta,* **172,** 277 (1969).
78. Heath, R. L., and G. Hind. *Biochim. Biophys. Acta,* **172,** 290 (1969).

CHAPTER 9

1. Izawa, S., T. N. Connolly, G. D. Winget, and N. E. Good. *Brookhaven Symposium, No. 19* (1967), p. 169.
2. Del Campo, F. F., J. M. Ramirez, and D. I. Arnon. *J. Biol. Chem.,* **243,** 2805 (1968).
3. Trebst, A., and E. Pistorius. *Biochim. Biophys. Acta,* **131,** 580 (1967).
4. Shavit, N., and V. Shoshan. *FEBS Letters,* **14,** 265 (1971).
5. Böhme, H., and A. Trebst. *Biochim. Biophys. Acta,* **180,** 137 (1969).
6. Nakamoto, T., D. W. Krogmann, and B. Mayne. *J. Biol. Chem.,* **235,** 1843 (1960).
7. Krogmann, D. W., and E. Olivero. *J. Biol. Chem.,* **237,** 3292 (1962).
8. Hauska, G. A., R. E. McCarty, R. J. Berzborn, and E. Racker. *J. Biol. Chem.,* **246,** 3524 (1971).
9. Brand, J., T. Baszynski, F. L. Crane, and D. W. Krogmann. *J. Biol. Chem.,* **247,** 2814 (1972).
10. Pratt, L. H., and N. I. Bishop. *Biochim. Biophys. Acta,* **153,** 664 (1968).
11. Cramer, W. A., and W. L. Butler. *Biochim. Biophys. Acta,* **143,** 332 (1967).
12. Lee, S. S., A. M. Young, and D. W. Krogmann. *Biochim. Biophys. Acta,* **180,** 130 (1969).
13. Black, C. C. Jr. *Biochem. Biophys. Res. Commun.,* **28,** 985 (1967).
14. Laber, L. J., and C. C. Black, Jr. *J. Biol. Chem.,* **244,** 3463 (1969).
15. Krogmann, D. W., and M. L. Stiller. *Biochem. Biophys. Res. Commun.,* **7,** 46 (1962).
16. Black, C. C., A. San Pietro, G. Norris, and D. Limbach. *Plant Physiol.,* **39,** 279 (1964).
17. Maclean, F. I., Y. Fujita, H. S. Forrest, and J. Myers. *Plant Physiol.,* **41,** 774 (1966).

18. Forti, G., and B. Parisi. *Biochim. Biophys. Acta,* **71**, 1 (1963).
19. Rurainski, H. J., J. Randles, and G. E. Hoch. *Biochim. Biophys. Acta,* **205**, 254 (1970).
20. Simmonis, W., in *Currents in Photosynthesis,* J. B. Thomas and J. C. Goodheer (eds.). Rotterdam: Donker, 1966, p. 217.
21. Kandler, O., and W. Tanner. *Ber. Deut. Botan. Ges.,* **79**, 48 (1966).
22. Jagendorf, A. T. *Fed. Proc.,* **26**, 1361 (1967).
23. Mitchell, P. *Fed. Proc.,* **26**, 1370 (1967).
24. Jagendorf, A. T., and J. Neumann. *J. Biol. Chem.,* **240**, 3210 (1965).
25. Izawa, S., and G. Hind. *Biochim. Biophys. Acta,* **143**, 377 (1967).
26. Jagendorf, A. T., and E. Uribe. *Proc. Natl. Acad. Sci. USA,* **55**, 170 (1966).
27. Schwartz, M. *Nature,* **219**, 915 (1968).
28. Nishimura, M., K. Kadota, and B. Chance. *Arch. Biochem. Biophys.,* **125**, 308 (1968).
29. Karlish, S. J. D., and M. Avron. *Biochim. Biophys. Acta,* **153**, 878 (1968).
30. McCarty, R. *Biochim. Biophys. Res. Commun.,* **32**, 37 (1968).
31. Nelson, N., Z. Drechsler, and J. Neumann. *J. Biol. Chem.,* **245**, 143 (1970).
32. McEvoy, F. A., and W. S. Lynn. *FEBS Letters,* **10**, 299 (1970).
33. Thore, A., D. L. Keister, N. Shavit, A. San Pietro. *Biochem.,* **7**, 3499 (1968).
34. Avron, M. *Biochim. Biophys. Acta,* **77**, 699 (1963).
35. McCarty, R. E., and E. Racker. *J. Biol. Chem.,* **243**, 129 (1968).
36. Farron, F., and E. Racker. *Biochem.,* **9**, 3829 (1970).
37. McCarty, R. E., J. S. Fuhrman, and Y. Tsuchiya. *Proc. Natl. Acad. Sci. USA,* **68**, 2522 (1971).
38. Ryrie, I. J., and A. T. Jagendorf. *J. Biol. Chem.,* **246**, 3771 (1971).
39. Karu, A. E., and E. N. Moudrianakis. *Arch. Biochem. Biophys.,* **129**, 655 (1969).
40. Evans, M. C. E., B. B. Buchanan, and D. I. Arnon. *Proc. Natl. Acad. Sci. USA,* **55**, 928 (1966).
41. Izawa, S., and N. E. Good. *Biochim. Biophys. Acta,* **102**, 20 (1965).
42. Katoh, S., and A. San Pietro. *Arch. Biochem. Biophys.,* **122**, 144 (1967).
43. Yamashita, T., and W. L. Butler. *Plant Physiol.,* **43**, 1978 (1968).
44. Inoué, H., and M. Nishimura. *Plant and Cell Physiol.,* **12**, 137 (1971).
45. Yamashita, T., and W. L. Butler. *Plant Physiol.,* **43**, 2037 (1968).
46. Selman, B. R., and T. T. Bannister. *Biochim. Biophys. Acta,* **253**, 428 (1971).
47. Mantai, K. E., J. Wong, and N. I. Bishop. *Biochim. Biophys. Acta,* **197**, 257 (1970).
48. Cheniae, G. M., and I. F. Martin. *Plant Physiol.,* **47**, 568 (1971).
49. Kimimura, M., S. Katoh, I. Ikegami, and A. Takamiya. *Biochim. Biophys. Acta,* **234**, 92 (1971).
50. Homann, P. H. *Biochim. Biophys. Acta,* **245**, 129 (1971).
51. Katoh, S., and A. Takamiya. *Plant and Cell Physiol.,* **12**, 479 (1971).
52. Epel, B. L., and R. P. Levine. *Biochim. Biophys. Acta,* **226**, 154 (1971).
53. Ben-Hayyim, G., and M. Avron. *Biochim. Biophys. Acta,* **205**, 86 (1970).
54. Inoué, H., and M. Nishimura. *Plant and Cell Physiol.,* **12**, 739 (1971).
55. Ben-Hayyim, G., and M. Avron. *Eur. J. Biochem.,* **15**, 155 (1970).

56. Heath, R. L. *Biochim. Biophys. Acta,* **245**, 160 (1971).
57. Hind, G., H. Y. Nakatani, and S. Izawa. *Biochim. Biophys. Acta,* **172**, 277 (1969).
58. Vernon, L. P., and E. R. Shaw. *Biochem. Biophys. Res. Commun.,* **36**, 878 (1969).
59. Honeycutt, R. C., and D. W. Krogmann. *Plant Physiol.,* **49**, 376 (1972).
60. Berzborn, R. *Z. Naturforsch.,* **23b**, 1096 (1968).
61. Brand, J., T. Baszynski, F. L. Crane, and D. W. Krogmann. *J. Biol. Chem.,* **247**, 2814 (1972).
62. Boardman, N. K., and J. M. Anderson. *Biochim. Biophys. Acta,* **143**, 187 (1967).
63. Park, R. B., and P. V. Sane. *Ann. Rev. Plant Physiol.,* **22**, 395 (1971).
64. Ogawa, T., L. P. Vernon, and H. H. Mollenhauer. *Biochim. Biophys. Acta,* **172**, 216 (1969).
65. Briantais, J. M. *Biochim. Biophys. Acta,* **143**, 650 (1967).
66. Arntzen, C. J., R. A. Dilley, G. A. Peters, and E. R. Shaw. *Biochim. Biophys. Acta,* **256**, 85 (1972).
67. Arntzen, C. J., R. A. Dilley, and F. L. Crane. *J. Cell Biol.,* **43**, 16 (1969).

CHAPTER 10

1. Kirk, J. T. O. *Ann. Rev. Plant Physiol.,* **21**, 11 (1970).
2. Kahn, A. *Plant Physiol.,* **43**, 1781 (1968).
3. Bogorad, L., in *Harvesting the Sun*, A. San Pietro, F. Greer, and T. Army (eds.). New York: Academic Press, 1967, p. 191.
4. Steer, B. T., and M. Gibbs. *Plant Physiol.,* **44**, 775 (1969).
5. ———. *Plant Physiol.,* **44**, 781 (1969).
6. Chen, S., D. McMahon, and L. Bogorad. *Plant Physiol.,* **42**, 1 (1967).
7. McMahon, D., and L. Bogorad. *Plant Physiol.,* **43**, 188 (1968).
8. Keller, C. J., and R. C. Huffaker. *Plant Physiol.,* **42**, 1277 (1967).
9. Margulies, M. M. *Biochem. Biophys. Res. Commun.,* **44**, 539 (1971).
10. Criddle, R. S., B. Dau, G. E. Kleinkopf, and R. C. Huffaker. *Biochem. Biophys. Res. Commun.,* **41**, 621 (1970).
11. Scott, N. S., H. Nair, and R. M. Smillie. *Plant Physiol.,* **47**, 385 (1971).
12. Filner, B., and A. O. Klein. *Plant Physiol.,* **43**, 1587 (1968).
13. Thorne, S. W., and N. K. Boardman. *Plant Physiol.,* **47**, 252 (1971).
14. Hillier, R. G., and N. K. Boardman. *Biochim. Biophys. Acta,* **253**, 449 (1971).
15. DeGreef, J., W. L. Butler, and T. F. Roth. *Plant Physiol.,* **47**, 457 (1971).
16. Oelge-Karow, H., and W. L. Butler. *Plant Physiol.,* **48**, 621 (1971).
17. Horak, A., and R. D. Hill. *Can. J. Biochem.,* **49**, 207 (1971).
18. Lockshin, A., R. H. Falk, L. Bogorad, and C. L. F. Woodcock. *Biochim. Biophys. Acta,* **226**, 366 (1971).
19. Forger, J. M., III, and L. Bogorad. *Biochim. Biophys. Acta,* **226**, 383 (1971).
20. App, A. A., and A. T. Jagendorf. *Biochim. Biophys. Acta,* **76**, 286 (1963).

21. Kirk, J. T. O., and M. J. Keylock. *Biochem. Biophys. Res. Commun.*, **28**, 927 (1967).
22. Harris, R. C., and J. T. O. Kirk. *Biochem. J.*, **106**, 34p (1968).
23. _____ . *Biochem. J.*, **113**, 195 (1969).
24. Schiff, J. A., M. H. Zeldin, and J. Rubman. *Plant Physiol.*, **42**, 1716 (1967).
25. Brown, R. D., and R. Haselkorn. *Proc. Natl. Acad. Sci. USA*, **68**, 2536 (1971).
26. Perl, M. *Biochem. J.*, **125**, 401 (1971).
27. Trémoliéres, A., and M. Lepage. *Plant Physiol.*, **47**, 329 (1971).
28. Hoober, J. K., P. Siekivitz, and G. E. Palade. *J. Biol. Chem.*, **244**, 2621 (1969).
29. Eytan, G., and I. Ohad. *J. Biol. Chem.*, **247**, 112 (1972).
30. Knight, G. J., and C. A. Price. *Biochim. Biophys. Acta*, **158**, 283 (1968).
31. Ridley, S. M., and R. M. Leech. *Nature*, **227**, 463 (1970).
32. Kirk, J. T. O., and R. A. E. Tilney Basset. *The Plastids*. San Francisco: W. H. Freeman and Co., 1967.
33. Tewari, K. K. *Ann. Rev. Plant Physiol.*, **22**, 141 (1971).
34. Whitfeld, P. R., and D. Spencer. *Biochim. Biophys. Acta*, **157**, 333 (1968).
35. Kung, S. D., and J. P. Williams. *Biochim. Biophys. Acta*, **195**, 434 (1969).
36. Chiang, K. S., and N. Sueoka. *Proc. Natl. Acad. Sci. USA*, **57**, 1506 (1967).
37. Tewari, K. K., and S. G. Wildman. *Symp. Soc. Exp. Biol.*, **24**, 147 (1970).
38. Wong, F. Y., and S. G. Wildman. *Biochim. Biophys. Acta*, **259**, 5 (1972).
39. Wells, R., and R. Sager. *J. Mol. Biol.*, **58**, 611 (1971).
40. Iwamura, T., and S. Kuwashima. *Biochim. Biophys. Acta*, **174**, 330 (1969).
41. Vedel, F., F. Ouetier, M. Byen, A. Rode, and J. Dalmon. *Biochem. Biophys. Res. Commun.*, **46**, 972 (1972).
42. Stutz, E., and J. P. Vandrey. *FEBS Letters*, **17**, 277 (1971).
43. Manning, J. E., D. R. Wolstenholme, R. S. Ryan, J. A. Hunter, and O. C. Richards. *Proc. Natl. Acad. Sci. USA*, **68**, 1169 (1971).
44. Spencer, D., and P. R. Whitfeld. *Biochem. Biophys. Res. Commun.*, **28**, 538 (1967).
45. Tewari, K. K., and S. G. Wildman. *Proc. Natl. Acad. Sci. USA*, **58**, 689 (1967).
46. Spencer, D., and P. R. Whitfeld. *Arch. Biochem. Biophys.*, **132**, 477 (1969).
47. Polya, G. M., and A. T. Jagendorf. *Arch. Biochem. Biophys.*, **146**, 635 (1971).
48. Bottomley, W., H. J. Smith, and L. Bogorad. *Proc. Natl. Acad. Sci. USA*, **68**, 2412 (1971).
49. Tewari, K. K., and S. G. Wildman. *Biochim. Biophys. Acta*, **186**, 358 (1969).
50. Armstrong, J. J., S. J. Surzycki, B. Moll, and R. P. Levine. *Biochem.*, **10**, 692 (1971).
51. Bottomley, W., D. Spencer, A. M. Wheeler, and P. R. Whitfeld. *Arch. Biochem. Biophys.*, **143**, 269 (1971).
52. Hadziyev, D., and S. Zalik. *Biochem. J.*, **116**, 111 (1970).
53. Tewari, K. K., and S. G. Wildman. *Proc. Natl. Acad. Sci. USA*, **59**, 569 (1968).

54. Jaworski, A., P. R. Whitfeld, and A. Seigel. *Fed. Proc.,* **28**, 864 (1969).
55. Gillham, N. W., J. E. Boynton, and B. Burkholder. *Proc. Natl. Acad. Sci. USA,* **67**, 1026 (1970).
56. Boynton, J. E., N. W. Gillham, and B. Burkholder. *Proc. Natl. Acad. Sci. USA,* **67**, 1505 (1970).
57. Stutz, E., and H. Noll. *Proc. Natl. Acad. Sci. USA,* **57**, 774 (1967).
58. Payne, P. I., and T. A. Dyer. *Biochem. J.,* **124**, 83 (1971).
59. ———. *Nature, New Biol.,* **235**, 145 (1972).
60. Ingle, J. *Plant Physiol.,* **43**, 1448 (1968).
61. Thomas, J. R. *Fed. Proc.,* **31**, 914 Abs. (1972).
62. Surzycki, S. J., and J. D. Rochaix. *J. Mol. Biol.,* **62**, 89 (1971).
63. Lyttleton, J. W. *Biochim. Biophys. Acta,* **154**, 145 (1968).
64. Odintsova, M. S., and N. P. Yurina. *J. Mol. Biol.,* **40**, 503 (1969).
65. Vasconcelos, A. C. L., and L. Bogorad. *Biochim. Biophys. Acta,* **228**, 492 (1971).
66. Gualerzi, C., and P. Cammarano. *Biochim. Biophys. Acta,* **199**, 203 (1970).
67. Bogorad, L., in *Control Mechanisms in Developmental Processes*, M. Locke (ed.). New York: Academic Press, 1967, p. 1.
68. Gnanam, A., and J. S. Kahn. *Biochim. Biophys. Acta,* **142**, 493 (1967).
69. Zeldin, M. H., and J. A. Schiff. *Plant Physiol.,* **42**, 922 (1967).
70. Dyer, T. A., and R. M. Leech. *Biochem. J.,* **106**, 689 (1968).
71. Leis, J. P., and E. B. Keller. *Biochem.,* **10**, 889 (1971).
72. Burkhard, G., B. Eclancher, and J. H. Weil. *FEBS Letters,* **4**, 285 (1969).
73. Williams, G. R., and A. S. Williams. *Biochem. Biophys. Res. Commun.,* **39**, 858 (1970).
74. Barnett, W. E., C. V. Pennington, Jr., and S. A. Fairfield. *Proc. Natl. Acad. Sci. USA,* **63**, 1261 (1969).
75. Burkard, G., P. Guillemaut, and J. H. Weil. *Biochim. Biophys. Acta,* **224**, 184 (1970).
76. Kislev, N., M. Selsky, C. Norton, and J. M. Eisenstadt. *Fed. Proc.,* **31**, 874 Abs. (1972).
77. Reger, B. J., S. A. Fairfield, J. L. Epler, and W. E. Barnett. *Proc. Natl. Acad. Sci. USA,* **67**, 1207 (1970).
78. Bartels, P. G., K. Matsuda, A. Siegel, and T. E. Weier. *Plant Physiol.,* **42**, 736 (1967).
79. Ingle, J. *Plant Physiol.,* **43**, 1850 (1968).
80. Margulies, M. M. *Biochem. Biophys. Res. Commun.,* **44**, 539 (1971).
81. Drlica, K. A., and C. A. Knight. *J. Mol. Biol.,* **61**, 629 (1971).
82. Schrader, L. E., L. Beevers, and R. H. Hageman. *Biochem. Biophys. Res. Commun.,* **26**, 14 (1967).
83. Smillie, R. M., D. Graham, M. R. Dwyer, A. Grieve, and N. F. Tobin. *Biochem. Biophys. Res. Commun.,* **28**, 604 (1967).
84. Linnane, A. W., and P. R. Stewart. *Biochem. Biophys. Res. Commun.,* **27**, 511 (1967).
85. Aaronson, S., B. B. Ellenbogen, L. K. Yellen, and S. H. Hutner. *Biochem. Biophys. Res. Commun.,* **27**, 535 (1967).
86. Richards, O. C., R. S. Ryan, and J. E. Manning. *Biochim. Biophys. Acta,* **238**, 190 (1971).

87. Anderson, L. A., and R. M. Smillie. *Biochem. Biophys. Res. Commun.*, **23**, 535 (1966).
88. Ellis, R. J. *Science*, **163**, 477 (1969).
89. _____ . *Biochem. J.*, **116**, 28p (1970).
90. Ellis, R. J., and M. R. Hartley. *Nature, New Biol.*, **233**, 193 (1971).
91. Pigott, G. H., and N. G. Carr. *Science*, **175**, 1259 (1972).

CHAPTER 11

1. Furuya, M., in *Progress in Phytochemistry*, L. Reinhold, and Y. Liwschitz (eds.). London: Interscience Publishers, 1968, 1, 347.
2. Tanada, T. *Proc. Natl. Acad. Sci. USA*, **59**, 376 (1968).
3. Jaffe, M. J. *Science*, **162**, 1016 (1968).
4. Galston, A. W. *Proc. Natl. Acad. Sci. USA*, **61**, 454 (1968).
5. Haupt, W. *Physiol. Veg.*, **8**, 551 (1970).
6. Pratt, L. H., and R. A. Coleman. *Proc. Natl. Acad. Sci. USA*, **68**, 2431 (1971).
7. Rubinstein, B., K. S. Drury, and R. B. Park. *Plant Physiol.*, **44**, 105 (1969).
8. Roux, S. J., and W. S. Hillman. *Arch. Biochem. Biophys.*, **131**, 423 (1969).
9. Walker, T. S., and J. L. Bailey. *Biochem. J.*, **120**, 607 (1970).
10. Gardner, G., C. S. Pike, H. V. Rice, and W. R. Briggs. *Plant Physiol.*, **48**, 686 (1971).
11. Fry, K. T., and F. E. Mumford. *Biochem. Biophys. Res. Commun.*, **45**, 1466 (1971).
12. Kroes, H. H. *Biochem. Biophys. Res. Commun.*, **31**, 877 (1968).
13. Anderson, G. R., E. L. Jenner, and F. E. Mumford. *Biochem.*, **8**, 1182 (1969).
14. Mumford, F. E., and E. L. Jenner. *Biochem.*, **10**, 98 (1970).
15. Siegelman, H. W., B. C. Turner, and S. B. Hendricks. *Plant Physiol.*, **41**, 1289 (1966).
16. Linschitz, H., V. Kasche, W. L. Butler, and H. W. Siegelman. *J. Biol. Chem.*, **241**, 3395 (1966).
17. Taylor, A. O., and B. A. Bonner. *Plant Physiol.*, **42**, 762 (1967).
18. Scheibe, J. *Science*, **176**, 1037 (1972).
19. Oelze-Karow, H., P. Schopfer, and H. Mohr. *Proc. Natl. Acad. Sci. USA*, **65**, 51 (1970).
20. Ahmed, S. I., and T. Swain. *Phytochem.*, **9**, 2287 (1970).
21. Bellini, E., and W. S. Hillman. *Plant Physiol.*, **47**, 668 (1971).
22. Steward, F. C., and A. D. Krikorian. *Plants, Chemicals and Growth*. New York: Academic Press, 1971.
23. Whitmore, F. W. *Plant Physiol.*, **47**, 169 (1971).
24. Cheng, T. Y., and G. L. Hagen. *Biochim. Biophys. Acta*, **228**, 503 (1971).
25. Aghar, S., and C. R. K. Murti. *Biochem. Biophys. Res. Commun.*, **43**, 58 (1971).
26. Wareing, P. F., and G. Ryback. *Endeavour*, **29**, 84 (1970).
27. Milborrow, B. V., and R. C. Noddle. *Biochem. J.*, **119**, 727 (1970).
28. Lang, A. *Ann. Rev. Plant Physiol.*, **21**, 537 (1970).

29. Key, J. L. *Ann. Rev. Plant Physiol.,* **21**, 449 (1970).
30. Momotani, Y., and J. Kato. *Plant and Cell Physiol.,* **12**, 405 (1971).
31. Evins, W. H., and J. E. Varner. *Proc. Natl. Acad. Sci. USA,* **68**, 1631 (1971).
32. Johnson, K. D., and H. Kende. *Proc. Natl. Acad. Sci. USA,* **68**, 2674 (1971).
33. Skoog, F., and D. J. Armstrong. *Ann. Rev. Plant Physiol.,* **21**, 359 (1970).
34. Hecht, S. M., N. J. Leonard, W. J. Burrows, D. J. Armstrong, F. Skoog, and J. Occolowitz. *Science,* **166**, 1272 (1969).
35. Burrows, W. J., D. J. Armstrong, M. Kaminek, F. Skoog, R. M. Bock, S. M. Hecht, L. G. Dammann, N. J. Leonard, and J. Occolowitz. *Biochem.,* **9**, 1867 (1970).
36. Dyson, W. H., C. M. Chen, S. N. Alam, R. H. Hall, C. I. Hong, and G. B. Chheda. *Science,* **170**, 328 (1970).
37. Berridge, M. V., R. K. Ralph, and D. S. Letham. *Biochem. J.,* **119**, 75 (1970).
38. Wood, H. N. *Proc. Natl. Acad. Sci. USA,* **67**, 1283 (1970).
39. Van Overbeek, J. *Science,* **152**, 721 (1966).
40. Pratt, H. K., and J. D. Goeschl. *Ann. Rev. Plant Physiol.,* **20**, 541 (1969).
41. Lieberman, M., and A. T. Kunishi. *Plant Physiol.,* **47**, 576 (1971).
42. Mapson, L. W., J. F. March, M. J. C. Rhodes, and L. S. C. Wooltorton. *Biochem. J.,* **117**, 473 (1970).
43. Hyodo, H., and S. F. Yang. *Plant Physiol.,* **47**, 765 (1971).
44. Zelitch, I. *Ann. Rev. Plant Physiol.,* **20**, 329 (1969).
45. Schweizer, C. J., and S. K. Ries. *Science,* **165**, 73 (1969).
46. Williams, E. G. *Nature,* **227**, 84 (1970).

INDEX